CLONING

of Frogs, Mice, and Other Animals

The University of Minnesota Press
gratefully acknowledges the publication assistance
provided by Mr. and Mrs. Richard G. Gray, Sr.

CLONING

of Frogs, Mice, and Other Animals

Revised Edition of
Cloning: A Biologist Reports

Robert Gilmore McKinnell
Professor of Genetics and Cell Biology
College of Biological Sciences
University of Minnesota

University of Minnesota Press, Minneapolis

Published by the University of Minnesota Press,
2037 University Avenue Southeast, Minneapolis MN 55414

Printed in the United States of America

Library of Congress Cataloging in Publication Data

McKinnell, Robert Gilmore.
  Cloning of frogs, mice, and other animals.

  Previous ed. published as: Cloning: a biologist reports.
1979
  Bibliography:p.
  Includes index.
  1. Cloning. 2. Cell nuclei—Transplantation.
3. Embryology, Experimental. I. Title.
QH442.2.M32 1985   596'.016   85-2541
ISBN 0-8166-6906-6

For Beverly, Nancy, Robert, and Susan:
As before, you are the reason why.

Preface

I was pleased to be asked by the University of Minnesota Press to revise my book *Cloning: A Biologist Reports* because many advances in cloning and related fields have been made since the book was published. They are important and should be included in any contemporary account of cloning.

A major development is the production of mice by nuclear transplantation. I anticipated this in the first edition, noting that two techniques were already at hand that would make this possible: micromanipulation, permitting surgical implantation of a donor nucleus into a recipient egg to produce a cloned mouse, and fusion of a donor cell (and its nucleus) to the mouse egg by means of an appropriately treated virus. These were reasonable prognostications because preliminary experiments suggested their practicality. Since that time, one laboratory reported the production of cloned mice by surgical implantation of a donor nucleus and another reported the production of cloned mice by fusing a donor cell (and its nucleus) with a recipient egg by means of a treated virus. There is some disagreement concerning the results obtained with nuclear transfer in mice. This is discussed in the section on mouse nuclear transfer in Chapter 5.

Large domestic animals have economic importance related to food supply. It comes as no surprise, then, that substantial strides were made in the amplification, by microsurgical intervention, of young sheep, horse, and cow embryos. Progress in cloning of large domestic animals is also reported in Chapter 5.

New sections have been added on naturally occurring and human-assisted cloning of microorganisms, molecules, and higher organisms. Perhaps knowledge of the existence of spontaneous clones in the natural world, as well as clones of various nonhuman organisms contrived by humans, will help to provide a perspective for those who make judgments about the ethics of human intervention in the reproductive process. Humans are animals, of course, and if reproductive technology makes possible the production of cloned fish, frogs, mice, and cattle, then it is perhaps timely to speculate about possible human clones. I speculate, but I do not advocate. My speculation includes a tentative outline of how to clone humans (see Chapter 5). If ethical decisions are to be made, surely knowledge of potential procedures will be helpful (see the Epilogue).

A number of references to original scientific articles appear in this edition. They need not be read to understand this account of cloning. They are provided to assure the nonspecialist reader that the book is based on scientific studies published in scientific journals. They also make it possible for readers to pursue the topics discussed. The journals and books cited are available at biological and medical libraries of many universities.

Errors continue to abound in accounts of cloning. I expanded my discussion of these errors (Chapter 1) because I believe they have the potential of engendering unwarranted fear of science and technology. I think there is little reason for disquiet because of cloning. As with the first edition, my purpose in writing this revision was to provide for the nonspecialist who is curious about cloning an unambiguous explanation, without introducing ideas that are misleading because they have been simplified. I hope I have succeeded.

It is my pleasure to acknowledge colleagues who have read and made comments on the revised manuscript. These include Dr. Marion Namenwirth, Dr. David Biesboer, Dr. Norman Kerr, and Dr. Alan G. Hunter. I discussed with Dr. Donald B. Lawrence the antiquity of asexually propagated plants and with Dr. Ernst C. Abbe mechanisms of plant apomixis. Errol D. Seppanen was helpful in detecting errors in the manuscript, as was Debra Kane.

Much of this second edition was prepared while I was a Royal Society Guest Research Fellow in the Nuffield Department of Pathology, John Radcliffe Hospital, University of Oxford, England. The fellowship was awarded to provide support for the study of the spread of cancer within the body. That spread, known as metastasis, is the most malignant aspect of cancer. I use genetically identical experimental animals for some of my metastasis research. I produce my genetically identical animals by cloning. Thus the Royal Society of London has assisted my cloning studies and, by that action, the revision of this book. I express my deepest appreciation to the Royal Society of London and to Dr. David Tarin, Nuffield Reader in Pathology at Oxford, for their generosity in supporting my research.

"That sweet city with her dreaming spires (that) needs not June for beauty's heightening," the magnificent Bodleian and the convenient Cairns, and the hospitality of colleagues at the John Radcliffe Hospital and Green College, stimulated and sustained my enthusiasm for this revision.

R.G.M.
Saint Paul, Minnesota

Contents

CLONING
of Frogs, Mice, and Other Animals

 Why a Discourse on Cloning? Of E. coli, Quaking Aspens, and Frogs. Humans Too?

No human has been cloned. But the technology for human cloning, at least a limited type of cloning, is here. Science makes rapid advances. A revolution in reproductive biology is now taking place, that provides the technical means for cloning humans. Although no human has yet been cloned, many frogs and salamanders have. So too have carrots and fish. Mice cloning has been reported. Cattle embryos have been fragmented surgically resulting in genetically identical progeny. The birth of babies resulting from *in vitro* fertilization witnesses to the efficacy of procedures first developed in animals and now applied to humans. It is my purpose in this book to chronicle the events that have led to the still emerging technology of *in vitro* fertilization because that technology is directly applicable to potential human cloning. Information about the current state of reproductive biology is essential for the development of informed opinion. Providing that information is the rationale for this book.

MISINFORMATION ABOUT CLONING

Concern about cloning remains in the media. Newspaper articles, magazine stories, books, television shows, and movies—as well as cartoons dealing with cloning—continue to bombard the public. Unfortunately, much of this information is incorrect. Inaccurate information and an understandable public concern about whether a human has or will be cloned, with all the ethical and moral questions that raises, have resulted in a distorted view of what cloning is and why biologists choose to clone. The real story, thus far, may seem less dramatic; but in a way it is more heartening. It is an account not of the production of carbon-copy dictators, millionaires, and Einsteins, but of research that may provide solutions to the very human problems of food supply, cancer, and aging.

3

Let me provide several specific examples of inaccurate information and comment on its potential harm.

A seemingly trivial—but, in fact, critical—biological flaw in an ethical discourse was published by Paul Ramsey (1970). He asserted that cloning is "reproduction by enucleating and renucleating an egg that has already been launched into life by ordinary bi-sexual reproduction. The question, Shall we clone a man? means, Shall we renucleate the human fertilized egg?" Ramsey, a Professor of Religion at Princeton University, described the prototypic cloning experiment of Briggs and King at Philadelphia as transplanting nuclei into "freshly fertilized egg cells." Actually, frog cloning involves the enucleation of an *unfertilized* egg. And that makes all the difference. Why? First, there is a technical reason. In frog (and other amphibian) eggs, the maternal hereditary material of an *unfertilized* egg is physically accessible and therefore enucleatable. Enucleation of the host egg is an absolute prerequisite for cloning by nuclear transplantation (described in Chapter 3). The nucleus of a *fertilized* intact frog egg is not visible, even with the aid of the best microscope; therefore, surgical enucleation is not feasible. No one, to my knowledge, has ever successfully enucleated a fertilized frog egg. If it *were* possible, I would not be concerned. With human eggs, I would. And this is the second reason why it makes all the difference. Although I might consider examining, studying, manipulating, and dissecting a human egg untouched by human sperm, I would, in fact, be loath to contrive an experiment on an egg already "launched into life by ordinary bi-sexual reproduction." Such an egg could be obtained only from the oviducts (natural fertilization ordinarily occurs in the oviducts, a process referred to as fertilization *in fallopio*, De Cherney, 1983). Removal of a fertilized egg from the reproductive tract of a woman could be considered an abortion. To abort for purely experimental reasons is clearly unthinkable. Fertilization can occur in laboratory glassware, and this has become increasingly common in the 1980s. But Ramsey was not referring to test-tube, *in vitro*, fertilization, for surely this is not "ordinary bi-sexual reproduction." It seems to me that misconceptions of this nature engender apprehensiveness about contemporary biology that is not warranted.

In 1974 Joseph Fletcher, theologian and Professor of Medical Ethics at the University of Virginia School of Medicine, described

4

cloning in a similarly confusing fashion: "A fertilized ovum or zygote is extracted from the oviduct and the fertilizing done *in vitro*. Next, its nucleus is removed (enucleated) and a body or 'somatic' cell is donated" (Fletcher, 1974).

These modestly faulty descriptions of cloning were provided by theologians, not biologists, and they appeared some time ago. But faulty information about cloning persists even among biologists. An example: cloning of frogs is described as a process whereby "the fusion-nucleus was removed *after fertilization* (italics added) and nuclei from later stage cells were inserted in its place" (Grobstein, 1981). This recurring minor mistake in the description of frog cloning is curious in two ways: first, because it is so common (see the description of nuclear transplantation, p. 240, by Lederberg, 1972; and, p. 207, by Watson et al., 1983) and because it anticipated a procedure in cell surgery similar to the micromanipulations used in the nuclear transfer procedure in *mice* in 1981 (Chapter 5).

The cloning procedure has elicited much thoughtful comment from theologians and life scientists. Of course, popular writers have also discussed cloning. I would like to comment on a widely discussed book by a popular writer. A free-lance science writer asserted that a wealthy businessman had had himself cloned. The book, Rorvik's 1978 *In His Image, The Cloning of a Man*, contains factual errors of biology.

Rorvik acknowledges by implication that human cloning has a model from experimental biology—the model of frog cloning. He states that nuclei from "frog body cells," when microsurgically inserted into unfertilized eggs deprived of their own nuclei, produce a "whole new frog" (pages 48–49, Rorvik, 1978). If Mr. Rorvik used "frog" to mean "adult frog" and not embryo, tadpole, or something else, he is unequivocally wrong. No body cell from an adult frog has ever been transplanted successfully to yield another "whole new frog" (see Chapters 3 and 4 of this book). The biological premise expounded in Rorvik's book, the premise that cloning of multicellular animals is possible with nuclei obtained from adult body cells, is incorrect—at least with the biotechnology available at the time of this writing.

I am flattered to be the subject of a paragraph in the notes of "methods" (#15, p. 212, Rorvik, 1978), but I claim less for my tumor nuclear transplantation experiments than does Rorvik (p. 139). He wrote: "Amphibians had been successfully cloned from what had been identified as adenocarcinoma cells . . . ." Rorvik describes successful as the production of a "whole new frog." My colleagues and I reported obtaining only tadpoles (not frogs) when renal adenocarcinoma nuclei were inserted into enucleated ova (Chapter 4). Thus Rorvik is an unreliable source of information on cloning in general and an unreliable source of information about some specific cloning studies in which I personally participated.

A U.S. District Court judge ruled Rorvik's book a "fraud and a hoax." Why was a court involved? Because a British biologist, Derek Bromhall, became irate when he read in Rorvik's book descriptions of procedures that he, Bromhall, had developed while working on rabbits (Bromhall, 1975). Rorvik claimed that these procedures were used in the purported human cloning. He had written to Bromhall, seeking "the current, acknowledged, state-of-the-art, in mammalian nuclear transplantation" to be used in a book he was writing. The date of the letter was May 1977, five months after the alleged birth of the cloned human. Bromhall filed suit (Broad, 1981).

Rorvik (*Newsweek*, February 4, 1980) retorted that "the court entirely dismissed the defamation charge, encompassed in two counts of alleged libel, and also summarily dismissed the charge alleging infringement of copyright, declaring that the plaintiff simply does not have a claim for defamation." Although that may be true, Judge John P. Fullam ruled that "the cloning described in the book never took place" and that "all the characters mentioned in the book, other than the defendant Rorvik, have no real existence" (Broad, 1981). Perhaps even more revealing is the fact that the publisher of Rorvik's book, J. B. Lippincott, reportedly settled with Bromhall for $100,000, issued a letter of apology to him (*New Scientist*, 22 April 1982, p. 202), and publicly admitted that they now believe the book by Rorvik to be untrue (*The Sunday Times*, London, 11 April 1982).

When Rorvik's book appeared, it was taken seriously by many people. He was invited to appear on television shows and became the principal cause of a congressional hearing. A Congressional Subcommittee on Health and the Environment responded to the public's interest in Rorvik's claim of human cloning with hearings in Washington, D.C., in May 1978. The subcommittee's purpose was to determine if the claim was true and to look into the direction that research in cell and genetic biology was taking. Questions: Can a human be cloned now or in the future? Why did biologists undertake experiments that led to the uncritical acceptance by many of the unsupported claim that a human had been cloned? Several biologists testified before the subcommittee, but Mr. Rorvik twice failed to appear—the first time citing health reasons, the second an extended promotional tour for his book.

It is difficult to summarize the congressional hearings because of the number of questions raised, the complexity of these questions, and the many points of view that were represented. I think the testimony of the late Dr. Andre E. Hellegers of the Kennedy Institute for the Study of Human Reproduction and Bioethics, Georgetown University, reflects the attitude of many working biologists: "I think fundamentally the problem is that too many people believe that cloning is an end, namely, to production of an individual. Factually, cloning is a means. It is a means of cell study and an enormously important one" (Hearing before the Subcommittee on Health and the Environment, 1978, p. 90).

Public apprehensiveness about cloning could lead to reduced funding for cloning research or even to regulations outlawing the procedure. This could well impede research that may contribute to the solution of such major human problems as food production, cancer, and aging. Certainly hazardous procedures should be subject to appropriate constraints. But I hope that this book allays apprehensiveness and convinces readers that the nuclear transplantation procedure known as cloning has thus far never been used in an unethical manner. Neither is it likely to be used in an unethical manner by those who pioneered its development. This is not to say that the procedure could not be abused. I know of no human activity or contrivance that is not subject to abuse. It is my hope that fear of abuse, abuse which is extraordinarily unlikely, does not impede scientific research that has a significant impact on basic human needs.

WHAT IS CLONING?

Molecules, microbes, a bewildering array of plant and animal species have been cloned by humans and by nature. All these diverse

7

cloned entities have in common the fact that sex is not a part of the process of their generation. "Clone," derived from the Greek "klon," meaning twig or slip, refers to asexual reproduction, also known as vegetative reproduction. The word has been adopted in the vernacular and is commonly used when referring to people with nearly identical physical appearance or exceptionally similar behavior.

Although this volume is concerned with animal cloning, I believe it is instructive to consider, if only briefly, the cloning of molecules, cells, and plants. It is my hope that this discussion will dispell the confusion that sometimes arises because of the diversity of end results, the "clones" (ranging from DNA, cells, plants and plant communities, to cloned frogs and mice) as well as the diversity of methods used in cloning. In some cases, no "methods" are used at all. Cloning is spontaneous, a part of nature, and the property of the living organism in which it occurs.

Cloned molecules, cells, plants, and animals are all genetically identical copies produced without the intervention of the sexual process. Of course, one can understand frog cloning without any consideration of gene cloning. This statement is particularly convincing when one considers that frog cloning preceded gene cloning by many years. But, a consideration of gene and other forms of "cloning" may lead to an appreciation of the remarkable progress of human endeavors in understanding biology at various levels of complexity. Humans clone molecules, cells, and organisms, and the capacity to do this witnesses to the extraordinary technological progress in biology of the past few years. One hopes discoursing on various methods of asexual replication will provide insight into what sexual reproduction is and why its absence is significant.

CLONES OF E. COLI AND OF DNA

Recombinant DNA technology that yields clones of DNA is as much in the news as is reproductive biology technology that yields clones of animals. The cloned DNA does not lead to the carbon copies of frogs or mice that are the primary subject of this book. But a better understanding of the frog-cloning process will probably result from understanding recombinant technology and the exceptionally valuable gene products that can be produced by the technology.

The procedures of recombinant technology are described with cell cloning because the former is dependent upon the latter.

Bacteria may be grown in dishes containing a nutrient medium (as well, of course, as in any other places where growth conditions are appropriate). If the nutrients are adequate and other conditions, such as temperature, are appropriate in the culture dish, bacteria will flourish *in vitro* (the term *in vitro* simply refers to the fact that the organisms are being cultured in a dish, traditionally glass; however, most culture dishes are now plastic). A colony of many millions of bacteria may grow from a single bacterium. If all the bacterial cells are produced by ordinary cell division, and if there was but one progenitor cell, then all members of the colony are genetically identical (except for the occasional mutant), and the colony is known as a clone.

In recent years techniques have been developed for attaching fragments of human (or any species) genetic material (i.e., DNA) to some of the bacteria's DNA. The human or other species DNA is linked ("spliced") to the bacterial DNA with the help of bacterial enzymes. The hybrid DNA is introduced into treated bacterial cells. The cells divide and as they do so they faithfully replicate not only their own DNA but also the foreign DNA that has been spliced into the bacterial genetic material. The clone of bacteria, consisting of many millions of cells, has now made many millions of identical copies of the foreign DNA, hence the term "DNA cloning."

The procedure of DNA cloning is possible because there are bacterial enzymes (known as restriction endonucleases) which will cut bacterial DNA, as well as the DNA of any species, in such a manner that the cut DNA has "sticky ends." There are other enzymes which will heal the cut. They are known as ligases.

A human gut bacterium, *E. coli*, is used for these experiments. The *E. coli* chromosomal DNA has different physical properties from another kind of DNA in the bacteria known as plasmid DNA. This permits the bacterial chromosomal DNA to be separated from the smaller plasmid DNA by centrifugation. Plasmid DNA occurs in circular form. When treated with a specific restriction endonuclease, the plasmid circle is cut in one place and its DNA becomes linear with "sticky ends." DNA from human cells (or any other source), in the meantime, may be treated with the same restriction endo-

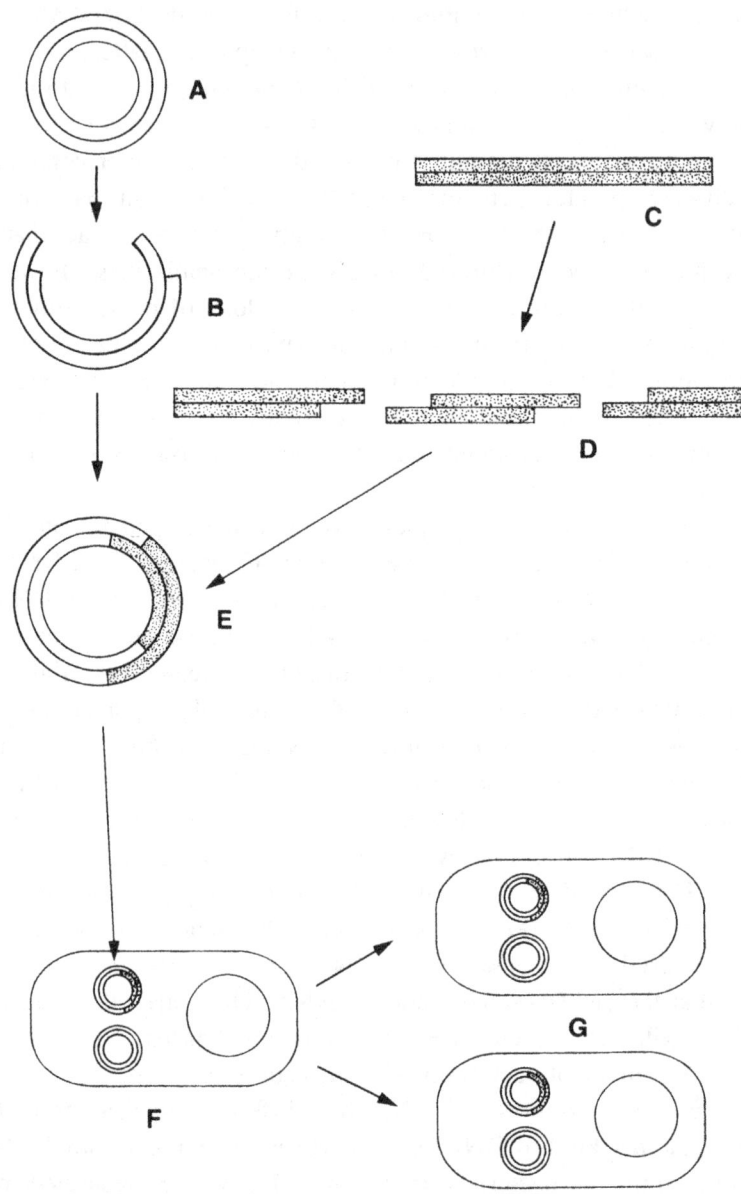

FIGURE 1-1. *Circular plasmid DNA (A), separated from bacterial chromosomal DNA by centrifugation, is converted into linear DNA with sticky ends (B) by means of a restriction endonuclease. Human, or other foreign DNA (C), is cut into fragments also with sticky ends (D), which can then be spliced to the plasmid DNA (E) by means of an enzyme known as a ligase, thereby forming a "chimeric" or hybrid plasmid. The hybrid plasmid can then be inserted into a properly treated bacterial cell (F), which then faithfully replicates ("clones") the synthetic plasmid (G).*

nuclease. Because the human DNA molecule is larger than the plasmid DNA, the enzyme cuts the human DNA into a number of fragments—each with sticky ends. The sticky human DNA fragments are then spliced with the bacterial plasmid DNA in the presence of a ligase. New circles of hybrid DNA (recombined) are formed. The hybrid DNA molecules (also known as DNA chimeras) are part bacterial plasmid and part human.

The hybrid DNA is inserted into previously prepared *E. coli*. The preparation consists of stripping the cell wall from the bacterial host leaving a naked protoplast. The naked protoplast readily admits the hybrid plasmid DNA. The newly engineered *E. coli* cell is cloned to give rise to a colony of genetically identical cells—each of which replicates not only its native DNA but the human DNA that was spliced to its plasmid DNA (Figure 1-1).

Bacterial clones can be grown that contain hybrid plasmid DNA with useful human DNA fragments. The human DNA can be provoked into directing the bacteria to produce a specific human gene product. An example of useful recombinant technology is the production of human insulin by appropriately engineered *E. coli*. There is an uncertain supply of animal-derived insulin for the increased demand by diabetic patients, some of whom become sensitive to animal-derived insulin. For these reasons, there is great potential value in insulin produced by recombinant DNA technology. What is more, huge numbers of *E. coli* containing human genes, can be grown relatively cheaply. Human insulin produced by *E. coli* is now available (Johnson, 1983).

The recombinant procedures, which have such promise not only for the production of human insulin but for other useful gene products as well, are described in many recent publications (for example, see Old and Primose, 1980; Gilbert and Villa-Komaroff, 1980; Hopwood, 1981; Maniatis et al., 1982; Watson et al., 1983).

It is a technical *tour de force* for humans to move DNA from one kind of cell to another; but nature has been doing just this sort of thing for a long time without help from humans. One such example involves the introduction of DNA from a bacterial virus, known as a phage, into a bacterial cell such as *E. coli*. When the phage DNA is introduced into the recipient cell, one of two things happens. Either the phage DNA becomes extremely virulent, causing the bac-

11

terial host to make many new phage bodies with the eventual rupture of the bacterium and release of phage, or, like the human recombinant procedure just described, the phage genetic material becomes spliced into the *E. coli* DNA and from then onward through many cell divisions the modified *E. coli* faithfully replicates ("clones") the phage DNA until, at some point, the viral DNA becomes virulent again and the host produces whole phage instead of just phage DNA.

Another example. Plants, such as tobacco, are susceptible to a neoplastic disease known as crown gall (Binns et al., 1981). The disease was thought to be induced by infection with a bacterium known as *Agrobacterium tumefaciens*. However, the bacterium is *not* needed for the continued growth of the tumor. It now appears that what the bacterium does is to introduce a bit of its plasmid DNA into the genetic material of the tobacco plant. The plasmid DNA becomes incorporated ("spliced") into the tobacco cell DNA and thereby conveys ("engineers") the tumorous state to the tobacco plant (Marx, 1979; Yadav et al., 1980; Caplan et al., 1983). Nontumorous traits can be selected and inserted into tobacco cells with the use of the *Agrobacterium* plasmid. Tobacco plants, regenerated from the treated cells, express the foreign gene and are fertile (Horsch et al., 1984).

CLONING PLANTS

House plants are easily propagated (cloned) from a twig or a slip; gardeners have been cloning potatoes for years. The edible part of the potato is an expanded stem known as a tuber, which, like other stems, has a number of buds or eyes. When placed in soil, each bud is capable of yielding an entire plant, and the crop so produced is a clone. This is an example of asexual reproduction. Plants usually reproduce sexually. The essential fact of sex in plants, or animals for that matter, is that hereditary material (DNA) from *two* individuals is joined to form a new creature. Just as each sexually produced organism varies in a diversity of subtle ways from its fellows of the same species, so too the sex cells provided by that unique plant or animal differ. It is not difficult to understand why each oak tree, or each rose blossom, is unique when one comprehends the nature of sex.

Pollen grains, which make some people sneeze in the summer, grow and form tubes when they find the sticky female portion of another flower. As the pollen grows (germinates) into the female part of the plant, two sperm nuclei move down the pollen tubes. One of the sperm nuclei fertilizes the egg and a new generation begins. The details of sex vary with different plants. The basic point is that single-celled plants, sea weeds, fungi, mosses, and all higher plants, including ferns, evergreens, and flowering plants, have the capability of reproducing sexually. Because sexual reproduction involves the union of genetically disparate sex cells, diversity is guaranteed.

Sexual reproduction is not involved in the cloning process. The cloned new plant does not result from the union of pollen-derived sperm with the ovum in a flower. The plant produced by cloning is a manifestation of the capacity for new growth and differentiation of the old plant body. Since no sexual reproduction is involved in propagation with a shoot or a twig, the new plant is genetically identical to the old plant. Cloning is used in agriculture to produce high-quality, uniform products. Desirable apple varieties are grown by grafting the variety onto an ordinary host tree. Seeds that result from sexual reproduction of the palatable variety of apple yield plants with the expected heterogeneity and fruit that is quite variable. Since non-uniform apples are not sought at the market, apples are cloned.

Apples and potatoes are not unique among vegetatively cultivated crops. Named varieties of grapes, edible varieties of bananas, sweet potatoes, sugar cane, pineapple, asparagus, and many other agriculturally important plants—even garlic—are "cloned." The cultivation of some of these plants has been known for at least 4,000 years (de Condolle, 1886).

Horticultural scientists are now developing culture methods for tiny fragments of economically important plants, but the biological principle is the same. Cloning is asexual reproduction whether it involves the growth of a relatively large twig or slip of a plant composed of many millions of cells, the growth of a small cluster of cells, or even the growth of a single cell.

Professor F. C. Steward and his associates at Cornell University in Ithaca, New York, produced carrots from the progeny of single cells and small cell clumps. Steward placed carrot cells from a mature plant on a culture plate containing nutrients and a gelatinlike

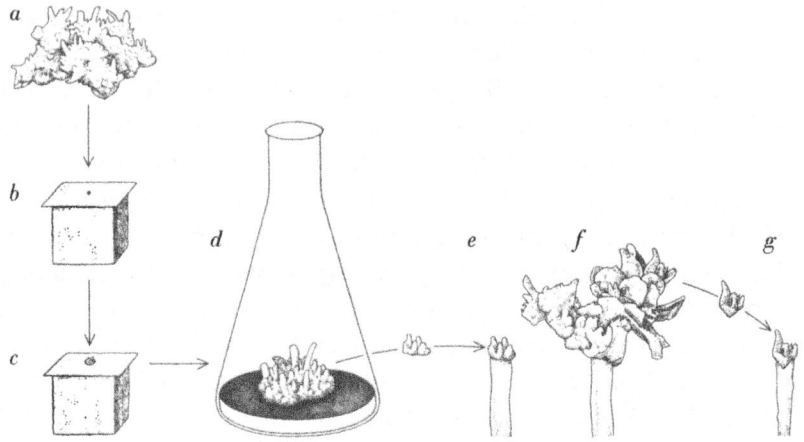

FIGURE 1-2. *Progeny of a plant tumor cell have the capacity to give rise to a normal plant. A tobacco tumor (a) provides a single cell (b) which is separated from nurse tissue by a filter. The cell proliferates into a small cluster (c) which is ex-* *planted to a culture flask (d). A leaf-like bud from the cultured tumor tissue is grafted to a normal tobacco host (e), where it develops more tumor tissue and other buds (f). A second bud is grafted to another tobacco plant host (g), where it*

material known as agar. A disorganized population of cells grew on the nutrient agar plate. Isolated free cells were removed from the mass and cultured in a liquid medium containing coconut milk, other nutrients, and plant hormones. Embryolike structures formed after repeated cell division in the liquid growth environment. Thousands of these structures developed roots and shoots and could be grown to mature carrot plants. What was accomplished with carrots was also effected in a number of other plant species including water parsnip, coriander, tobacco, and orchids. (For a more complete list of plant species that have been cloned, as well as procedures for clonal propagation, see Dodds and Roberts, 1982.) The vegetative propagation of single plant cells in the laboratory to produce carrots and parsnips is the ultimate in plant cloning—at least with regard to the size of the initial slip, which is, of course, microscopic (Steward, 1970; see also Backs-Hüsemann and Reinert, 1970; Haccius, 1978; Konar et al., 1972; Chaleff, 1983).

A surprising result in plant cloning was reported by Professor Armin Braun of the Rockefeller University of New York City. A plant tumor cell gave rise to a population of cells that eventually differentiated as normal tissue. Braun's experiment involved shaking a

develops extensive tissue (h), which pro-vides a third bud for grafting (i). That bud gives rise to normal tissue (j), flowers and seed (k). The seed will grow into a normal plant, but its cells fail to grow on a basic culture medium suitable for tumor cells.

(From A. C. Braun, "The reversal of tumor growth," Scientific American, November 1965, 76-77, Copyright © 1965 by Scientific American, Inc. All rights reserved)

plant tumor (tobacco teratoma) rapidly in culture so that single cells were obtained. A single cell was then cultured (cloned) until it formed a mass of tumor cells. Fragments of the mass were grafted to healthy plants. After repeated grafting, the tumor cells became progressively more normal in appearance until stems, leaves, and flowers formed. The flowers produced seed that, when sown, produced normal tobacco plants (Figure 1-2) (Braun, 1965; Braun and Wood, 1976; Binns et al., 1981). Braun's experiments are important for this discussion of cloning because they demonstrate vegetative reproduction—a single cell of a plant tumor is not a sex cell—and the reversibility of the malignant state (see Chapter 4 for another example).

Apomixis and Other Natural Plant Clones

Humility on the part of scientists is not inappropriate because, as noted in the discussion of DNA cloning, what humans often accomplish with great effort, nature does spontaneously. Apomixis is natural cloning that sometimes occurs in higher plants (Gustafsson, 1946). It is one kind of asexual reproduction. An example occurs in *Citrus* species where body cells (somatic diploid cells) form an em-

15

bryo directly, without union of sex cells (Frost, 1938). Because of this, the embryo is a genetic replicate of the parental plant—hence, cloned.

Other spontaneous clones occur on a grand scale. The quaking aspen (*Populus tremuloides*) develops from a seed, but early on the sapling sends out horizontal roots, many of which will give rise to shoots that eventually develop into trees. And the growing trees repeat the process of lateral root development to give rise to yet more shoots—until an enormous colony of genetically identical trees develops. One aspen clone was described as consisting of more than 47,000 trees covering an area of over 100 acres (Kemperman and Barnes, 1976; Cook, 1983).

Apomixis and the quaking aspen are examples of plant clones that develop without the intervention of humans. When the ethics of cloning experiments are considered later (Epilogue), it may be well to recall the spontaneous clones in nature. Further, the awe with which cloning scientists are sometimes viewed may be modulated when it is remembered that many of their efforts only mimic nature.

CLONING ANIMALS

Ordinarily, higher animals, like plants, reproduce sexually. However, there are experimental procedures that permit asexual reproduction in fish, frogs, salamanders, and mice. The procedures involve nuclear transplantation and are referred to as "cloning." Cloning could be referred to as apomixis or vegetative reproduction of animals. However, I suspect that few animal scientists would understand the word "apomixis," and it would be puzzling to the layperson to refer to a mouse or a frog as having been produced by *vegetative* reproduction.

Although several kinds of animals have been cloned, frogs have the distinction of being the *first* multicellular animals to be produced in this way. There are several reasons why. Frogs have an abundant supply of eggs and sperm that can be used by experimenters. A biologist may obtain from a single ovulation only 20 eggs from a mouse, but 2,000 from a frog (Figure 1-3). The fertilization and embryonic development of a frog, which ordinarily occurs outside the animal's body in ponds, is accomplished easily in the laboratory in glass dishes (*in vitro*). This permits direct observation of and experimenta-

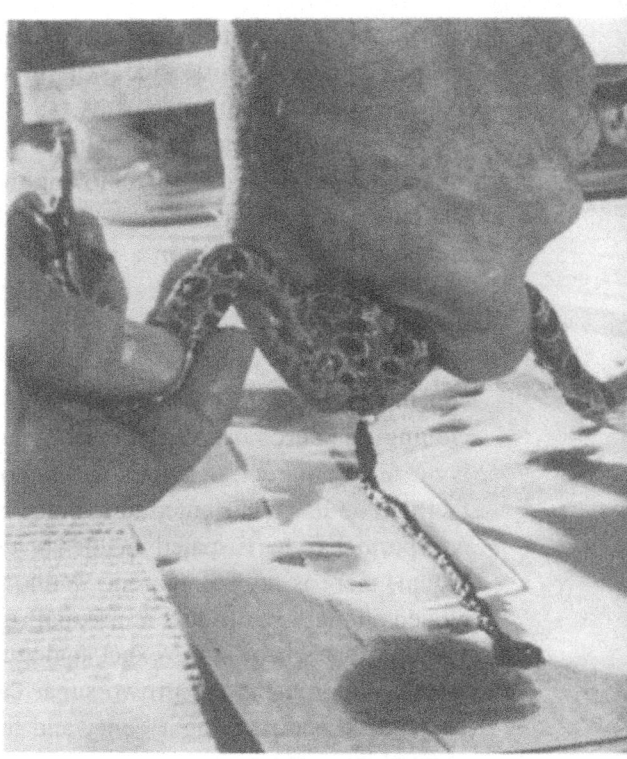

FIGURE 1-3. *Many eggs can be extruded from a gravid female frog by gentle pressure on the body wall.*

tion on all stages of development (in contrast to the fetal development of a mouse, which is hidden because the fetus develops in the uterus). The frog embryo matures into an organism with brain, eyes, liver, and other organs not unlike those of humans. The late nineteenth-century and early twentieth-century experiments that provided scientific groundwork for cloning were done with frogs and other amphibians. Thus history, a relevance to humans, as well as abundance of eggs and sperm and ease of handling embryos in the laboratory, were probably the principal reasons why frogs were cloned first.

Cloning frogs is not new. The microsurgical procedure of placing a frog nucleus into an egg deprived of its own hereditary material has been available to the scientific community for many years. Frogs have been asexually produced in some of the nuclear-insertion experiments. A cloned frog reveals that the donor nucleus had all the hereditary material (DNA) necessary for complete development. Perfection and use of the frog cloning procedure depended

17

on a sophisticated understanding of amphibian reproductive biology and an equally sophisticated skill in the manufacture and use of microscopic surgical instruments.

Successful nuclear transplantation in amphibians requires that the egg be enucleated, thus removing the maternal hereditary material contained in the egg nucleus. Then other hereditary material, contained in the nucleus from a body cell, is placed in the enucleated egg. The resulting cloned individual is parentless in the usual meaning of the word.

Biologically, a mother is a mother by virtue of the fact that she contributes hereditary material via the chromosomes of an egg. In cloning, the chromosomes of the egg have been enucleated, so there is no female parent. A father is a biological father by virtue of the fact that he has contributed hereditary material via his sperm. Since no sperm has participated in the development of the cloned individual, there is no male parent. Without a male or female parent, a cloned animal is a product of asexual reproduction. A frog produced by nuclear insertion is the exact analogue of a tree in an aspen clone, a stalk of commercially grown sugar cane—or a grafted apple. The frog, the aspen, the sugar cane, and the apple are all produced by asexual generation.

Why produce a frog asexually? Why climb Mount Everest? Mount Everest, it is said, is climbed "because it is there." Most scientists just don't work that way. They seek answers to substantial biological questions when they do a cloning experiment. It would be frivolous indeed to dissipate precious research funds to clone "because it is there." Frogs are not cloned to produce new frogs. It is certainly a more economic use of resources and time to let frogs reproduce sexually, as they have been doing so well for millennia. Scientists use the cloning procedure to gain insight into biological phenomena such as differentiation, cancer, immunobiology, and aging.

Is the hereditary material of an embryo or an adult cell equivalent to that of a fertilized egg? Is the hereditary material of a cancer cell or an aging cell the same as that of a fertilized egg? More crucial is whether the hereditary material of embryos, adults, cancer cells, and aging cells can be manipulated, coaxed, or provoked into expressing hereditary potentialities similar or equivalent to that of the

18

fertilized egg. More simply stated, is control of some of the most fundamental aspects of cell biology possible? ''Why clone a frog'' may be rephrased as ''Why try to understand differentiation?'' Cancer and aging are the most obvious areas where new understanding is needed. Many, including me, think that cloning may provide new and useful insight into these critically important biological problems. And cloning may even be helpful sometime in the future in overcoming rejection in organ transplantation.

Frogs were first cloned a third of a century ago. Mice were cloned more recently. What about human cloning? The procedures that would probably permit a special kind of human cloning are at hand. But human cloning is not apt to happen soon—if ever. Why not? There is no need to clone humans in order to provide answers to the questions that are central to cloning research. As I said, frogs are not cloned simply to produce new frogs. Human cloning would relate primarily to producing more people. Reproduction by cloning is an inappropriate means to reproduce more frogs, mice, or humans because species survive through genetic heterogeneity. Sexual reproduction ensures diversity, but there is sameness among individuals reproduced asexually. Most of us treasure uniqueness, especially among family and friends, and survival of the species (Stebbins, 1950; also see Cole, 1984) demands heterogeneity—not sameness. Thus cloning, although a useful procedure to the horticulturalist and experimental biologist, is not an appropriate method for human reproduction. It seems that 4 billion plus souls are enough for this fragile sphere. Cloners seek a better life for those who are already here.

Efforts to clone humans clearly are not in the mainstream of biological research. Human cloning attempts will be costly and difficult. Certainly no scientific question will be answered. However, human cloning presents a challenge, and some people respond to technical challenge just as others, like the mountain climber, respond to physical challenge. Therefore, it will not be surprising if occasionally someone attempts human cloning. I trust that that individual, should he or she ever be successful, will experience great personal gratification for having perfected a difficult procedure. But he or she will add little or nothing to the welfare of humans.

A theologian and some biologists make modest errors in biology—a science writer creates a concern that extends all the way to the Congress of the United States. James Dewey Watson, Nobel laureate in molecular biology, said, "It might be expected that many biologists, particularly those whose work impinges upon (cloning), would seriously ponder its implication, and begin a dialogue which would educate the world's citizens" (Watson, 1971).

This book is my contribution to that dialogue.

 **"A Fantastical Experiment"**

Fish, frogs, and mice have been produced by nuclear transplantation. With success in such diverse animals, it would seem that there is more than just the technical possibility that humans may some day be cloned. Indeed, we may be on the threshold of human cloning.

Since the cloning of humans raises enormous ethical and moral issues, and since it fails to heal any known disease, it seems appropriate to ask the question that was asked in a medical journal: "Should we have taken the first step?" (Genetic Engineering: Reprise. Editorial, *The Journal of the American Medical Association*, 220:1356–57, 1972). The first step in the present situation probably refers to the experimental studies that have made the cloning of a human being at least a theoretical possibility.

Designating any particular experiment as the *first* step is entirely arbitrary. This statement seems especially apt when one considers the extent to which cloning occurs in nature without human intervention (Chapter 1). However, several late nineteenth-century experiments were instrumental in providing new insight into the cell biology of development—and they ultimately provoked and enabled the nuclear-transplantation experiments. Thus, the first step(s) in animal cloning were taken years ago, and these studies are the substance of classical embryology—classical not because it is ancient, but because it led directly to the research programs of late twentieth-century cell and molecular biology. To ignore or bury these studies would be to obliterate the foundations of contemporary biomedical progress. These experiments hold the promise of new procedures that *may* enhance animal production for human food and *will* provide new insight into ancient human maladies such as cancer and aging.

THE "FIRST STEP(S)"

Nineteenth-century biologists sought to determine if cells of an embryo develop independently or if they interact and affect the fate of each other during embryogenesis. Is the early embryo a mosaic of autonomous cells each developing according to intrinsic control and producing a population of diverse and differentiated cell types that will comprise the adult, or is the early embryo a mass of cells that influence one another and that, by coordinated effort, produce an adult constituted of the various cell types? The early experiments designed to answer these questions relate not only to embryonic cells but also to cell components—nucleus and cytoplasm.

A fertilized egg, also known as a zygote, consists of a nucleus and cytoplasm (Figure 2-1). In fact, almost all cells—brain cells, liver cells, skin cells, as well as zygotes—are composed of nuclei and cytoplasms. Nuclei contain DNA (deoxyribonucleic acid), the hereditary or genetic material of the individual. The cytoplasms contain the biochemical and physiological support systems for the maintenance of the cell and the individual. One need not take a course in histology, the study of microscopic anatomy of tissues, to comprehend that the anatomy of a brain cell is different from that of a liver cell and the liver cell structure differs from that of a kidney cell, skin cell, etc. (Think of the differences in the texture and color of brain and liver in the supermarket). The term "differentiation" refers to the

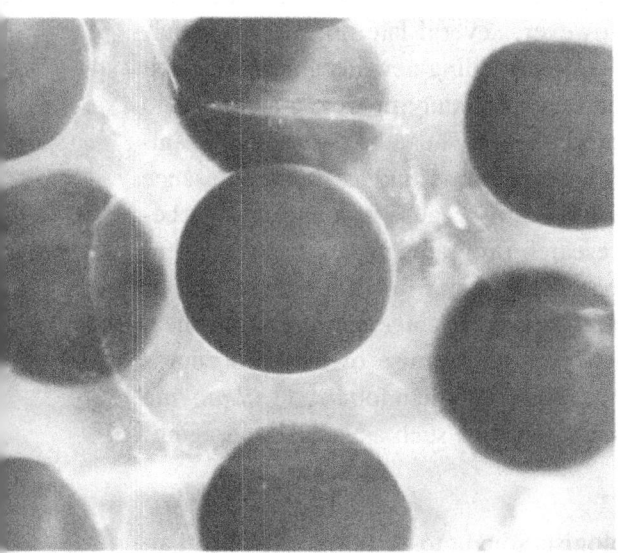

FIGURE 2-1. *Freshly fertilized eggs of the northern leopard frog,* Rana pipiens. *Each egg is contained in a transparent jelly membrane.*

developmental events that lead to these obvious differences, to the many cell changes that occur as development proceeds from the zygote to the mature organism. It is in the mature organism (and, of course, in *maturing* organisms) that component tissues manifest their unique structures and functions.

In human beings and other higher animals, the differentiated state is extraordinarily stable. If a person is fortunate enough to live to be 90, his or her liver remains liver, kidney remains kidney, skin is stabilized as skin. The extraordinary stability of the differentiated state, which we take for granted, was the subject of the investigations of pioneer biologists.

August Weismann (1834–1914), a German who taught zoology and comparative anatomy at Freiburg, published a landmark theory relating to the nature of the differentiated state. Weismann's theory is no longer tenable, but he has a memorable position in the history of embryology because of the questions he raised. Weismann theorized that the fertilized egg contains all the genetic determinants to form a complete individual. Thus the nucleus of a zygote, which contains the genetic material DNA, should have all the genetic material to form an entire individual, since an entire individual results from the development of the zygote (Weismann, 1892).

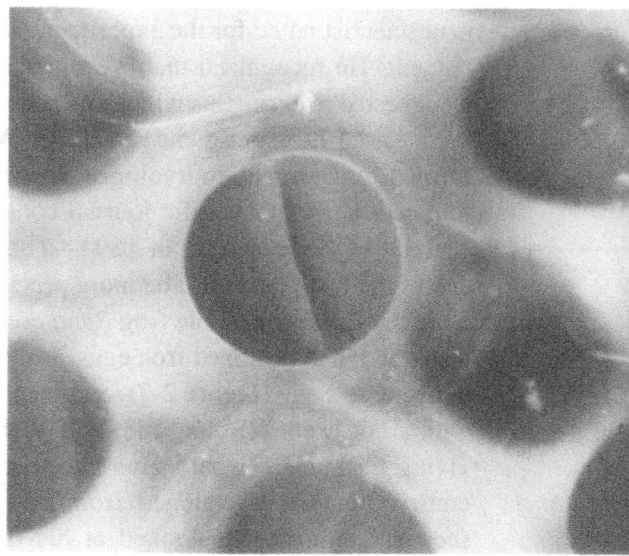

FIGURE 2-2. *Two-blastomere stage of the northern leopard frog,* Rana pipiens. *An egg divides into two blastomeres about 3.5 hours after fertilization at 18°C.*

The zygote first divides into two cells known as blastomeres (Figure 2-2). Blastomere, derived from Greek, means "part of a bud." The fertilized egg is the unopened bud of a new individual. When that bud divides into two, then four, then eight, then more cells on its way to flower as a new adult, the early cells are, of course, parts of the bud.

Weismann supposed that the genetic determinants of the zygote are divided when the egg divides. Thus the right blastomere of a two-cell stage would contain the genetic determinants (genes or DNA) to form the right half of an embryo, and the left blastomere would contain the genetic determinants to produce the left. When the two-blastomere stage gave rise to the four-blastomere stage, the genetic determinants would be divided so that each of the four cells would contain one-fourth of the DNA originally present in the zygote nucleus. This process would continue sequentially until the liver was structured with cells containing genetic determinants (DNA) only for liver, the brain-cell nuclei with genetic determinants only for brain, and skin nuclei DNA that is specific for skin. Weismann's hypothesis certainly seemed reasonable because of the extraordinary stability of the differentiated state. The person who lived to be 90 had liver that remained liver because the genetic determinants contained in liver cells could specify for *nothing but liver*—or so Weismann believed.

Hans Spemann, who will be discussed later, was a pioneer German scientist noted for the experimental analysis of amphibian development. He recognized that Weismann's theory was useful because it suggested experiments which could be performed in the laboratory.

An early experiment related to Weismann's theory was performed by German embryologist Wilhelm Roux (1850–1924), who founded the first scientific journal concerned with the experimental analysis of development in 1894—it is still published. Toward the end of the last century (to be more precise, during the spawning period of the European edible frog *Rana esculenta* in the spring of 1887), Roux obtained fertilized frog eggs, waited until they reached the two-blastomere stage (Figure 2-2), and then killed one of the blastomeres with a hot needle (Roux, 1888). If Weismann was correct, the surviving blastomere would give rise to only half an embryo. A half embryo, in fact, developed from this seminal experiment, and it seemed to those who looked at his illustrations (Figure 2-3) that

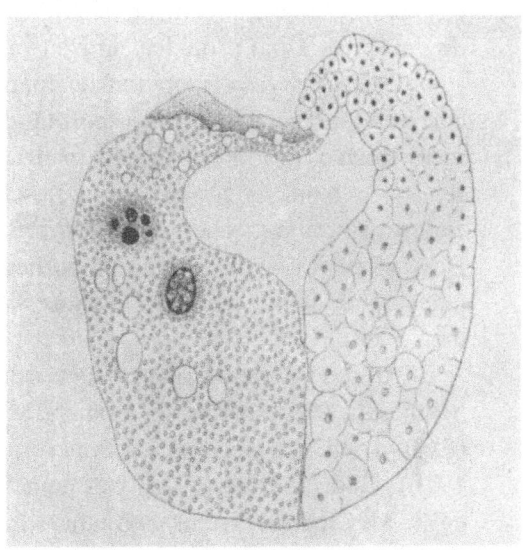

FIGURE 2-3. *A half embryo develops following the death of one of the first two blastomeres. The embryo part on the right is cellular; the mass to the left is the remains of the dead blastomere. (From Roux, 1888)*

Roux's experiment did, in fact, confirm Weismann's theory. I will say more about Roux's experiment later.

About the same time that Roux was toiling with frog eggs, German embryologist Hans Adolph Eduard Driesch (1867–1941) was experimenting with sea urchin eggs. Sea urchins, small marine creatures related to starfish, are to this day exceptionally important in studies of fertilization and developmental biology. Driesch, in effect, performed the same experiment that Roux did, only in a different way. Sea urchin eggs are much smaller than the eggs of amphibians, so it would have been difficult for Driesch to destroy a blastomere by inserting a hot needle into it. Instead, he took sea urchin eggs at the two-blastomere stage, placed them in a flask, and shook the flask so vigorously that the two cells became detached from each other.

Driesch carefully recorded what happened to the separated blastomeres. Each developed as a whole embryo but they were dwarfed. They were smaller than normal probably because the blastomeres were smaller than the initial undivided egg. But the most important event he observed was that they were anatomically whole embryos, not half embryos (Driesch, 1892). It would seem that the philosophical implications of obtaining an entirety from a fragment

25

were so astonishing that Driesch abandoned experimental embryology and became a professor of philosophy. Manipulative embryology still has a philosophical impact in the late twentieth century.

Driesch's results obviously contrasted with Roux's. Driesch correctly observed that one important difference was that the blastomere treated with a hot needle in Roux's frog experiment was not *separated* from its compatriot. Driesch thought that the separation experiment ought to be done with fertilized eggs of higher animals. He attempted it with frogs and mourned, "I have tried in vain to isolate amphibian blastomeres, let those who are more skillful than I try their luck."

Among those who were successful was the late University of Minnesota Professor Jesse Francis McClendon, who reported the development of isolated blastomeres of a frog egg (McClendon, 1910). Gudrun Ruud (1925) was equally successful with salamander eggs. McClendon, Ruud, and others reported that at the two-blastomere stage, under certain conditions, each of the blastomeres when completely separated from the other, would develop into a whole, intact embryo. Thus vertebrate embryos (vertebrates, animals with back bones, include frogs, toads, salamanders, as well as fish, reptiles, birds, and mammals including humans) seemed to have the same capacity to develop from separated blastomeres as did sea urchin embryos (see Chapter 5 for a discussion of blastomere separation in mice and large domestic animals). The conclusion we draw from these pioneer experiments is that, at least in the earliest part of development, genetic determinants are *not* divided among the blastomeres as Weismann postulated.

Let's return briefly to Roux's experiment involving the insertion of a hot needle into one blastomere at the two-cell stage of a frog embryo, with a half embryo resulting. Roux was a good reporter who failed to draw the right conclusions. The most accurate explanation of Roux's results seems to be that the mass of the dead blastomere, in intimate contact with the living cell progeny of the surviving blastomere, physically prevented movements of cells and otherwise inhibited the viable half from fully expressing its genetic potentialities. Blastomere *separation* is a better test of the developmental capabilities of a frog-embryo fragment.

Thus with sea urchins, frog embryos, and some mammals there

seems to be no diminution of genetic potentialities when the egg divides into blastomeres at an early stage. The nucleus of each blastomere contains *all* the genetic information needed to specify for a whole organism. However, an adult is composed of millions upon millions of cells. Is there any way to ascertain what the genetic capabilities are of nuclei obtained from the cells of older embryos or adults?

PRIMITIVE CLONING

The German-born American physiologist Jacques Loeb (1859–1924) studied parthenogenesis, the development of *unfertilized* eggs. He used various salt solutions to stimulate development of eggs that had not been fertilized by sperm. He noted that in some *fertilized* eggs subjected to this unnatural shock the cell membrane tears and egg cytoplasm protrudes. In effect, this is a herniated egg—a nucleated mass (nucleus and cytoplasm) with a small appendage of non-nucleated cytoplasm. The nucleated cytoplasm divides when the zygote nucleus divides. The appended bleb, bereft of a nucleus, fails to divide for a time. Eventually, however, a nucleus may move across the cytoplasmic bridge into the formerly non-nucleated bleb. This movement of the nucleus, described by Loeb while he was at the University of Chicago (1894), is a nuclear transplantation experiment, a *cloning experiment of nature* made possible by the altered seawater. Sometimes the cytoplasmic protrusion becomes another embryo after nucleation, a twin of the first embryo. The important biological message of this spontaneous nuclear transplantation is that the genetic determinants of the zygote nucleus are not partitioned during cleavage divisions of the sea urchin embryo, which would have happened had Weismann been correct. For if genetic determinants *had* been partitioned during cleavage, the late cleavage nucleus that entered the cytoplasmic bleb could not have contained the full complement of genetic information needed to direct development of a complete embryo.

Hans Spemann (1869–1941), Professor at Freiburg and Nobel laureate referred to earlier, was aware of Loeb's experiments and wanted to extend the observations to vertebrate embryos. His method involved the constriction of a salamander zygote with a noose made of baby hair. (Baby hair was a valuable commodity in embryology laboratories at the turn of the century because very fine monofilament line was not available and baby hair was the finest and strongest thread at hand). The zygote became dumbbell-shaped as the baby-hair noose was gently tightened. The salamander zygote nucleus was in one portion of the dumbbell-shaped egg. The other portion was nucleusless, not unlike the herniated cytoplasmic protrusion of the sea urchin embryo described by Loeb. The nucleated portion of the egg divided into two, then four, then eight, then sixteen cells. At about this time, Spemann gently loosened the noose sufficiently so that one of the nuclei could traverse the cytoplasmic bridge to the un-

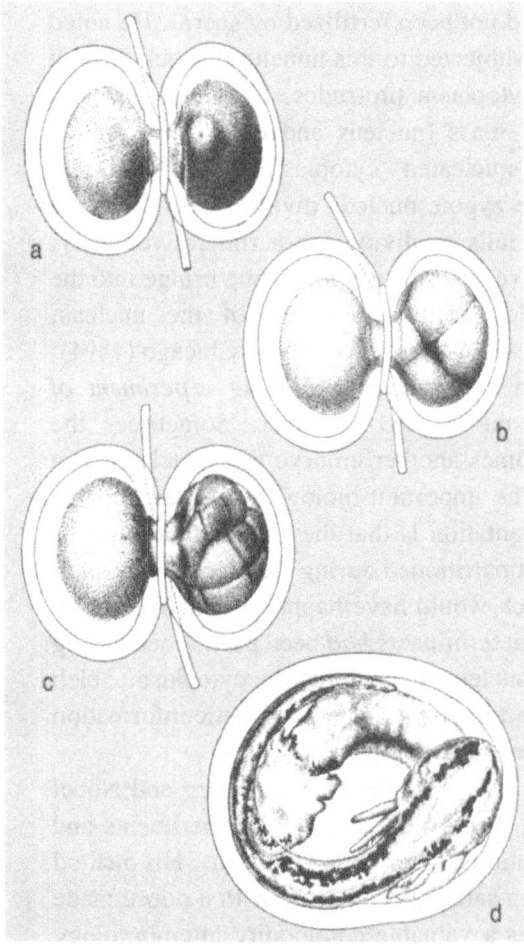

FIGURE 2-4. *Twin embryos developed when a fertilized egg was constricted with a loop of baby hair in an experiment designed by Hans Spemann to test the developmental equivalence of early embryonic nuclei. (a) A fertilized egg has been constricted with a loop of baby hair. Only the egg cytoplasm on the right of the loop contains a nucleus. (b) Cleavage into four blastomeres occurs in the nucleated portion of the constricted egg. (c) A nucleus is permitted to move into the previously non-nucleated portion of the egg cytoplasm to the left at about the 16-cell stage. Following this, the previously non-nucleated cytoplasm begins to divide with the result that (d) two whole and normal embryos result. The embryo to the upper left is somewhat younger than the embryo to the lower right because this embryo started its development several hours later. (From H. Spemann, "Die Entwicklung seitlicher und dorso-ventraler Keimhälften bei verzögerter Kernversogung," Zeitschrift für wissenschaftliche Zoologie, 1928, 132:105-34, Figures 17, 19, 21, 23)*

nucleated portion of the egg. When this occurred, that portion of the egg began to divide. Spemann then separated the two egg fragments completely (Spemann, 1938).

According to Weismann, the nucleus that invaded the egg fragment that was previously unnucleated would contain only a fraction of the genetic determinants of the zygote, and, therefore, the second embryo would develop incompletely. Instead, development was normal but somewhat delayed—delayed because the secondary embryo got a later start (Figure 2-4). The significance of this experiment is that both fragments developed as whole, complete embryos. Weismann was wrong.

Spemann contemplated these results and wished that he could place the nucleus from a more differentiated cell in the cytoplasm of an egg deprived of its own nucleus. He referred to this postulated manipulation as a "fantastical experiment." Spemann could envision the isolation of a nucleus; he could envision a fragment of egg cytoplasm devoid of a nucleus; at that time, however, he could see no way of introducing the isolated nucleus into the egg cytoplasm (Spemann, 1938).

TOOLS FOR CLONING

Although Spemann could foresee no way of inserting a nucleus into egg cytoplasm in 1938, within only 14 years, his "fantastical experiment" was a fait accompli, performed by Robert Briggs and Thomas J. King in Philadelphia. The apparatus necessary for cloning, needing only minimal modification for use in nuclear transplantation, had long been present when Spemann's book was published in 1938. Thus Spemann's frustrations were unwarranted.

Problem solving is always simpler retrospectively than prospectively. But how else in 1938 could a nucleus be placed in egg cytoplasm except by injection with a very fine diameter glass tube? (More recently, viruses have been used to assist in nuclear transplantation—but more of that in Chapter 5.) The injection of material into cells requires a glass tube that is small enough to penetrate the cell without permanently damaging it, but large enough to accommodate whatever is being injected. The injection procedure is enhanced if there is an instrument to grasp and to move the fine glass tube with precision.

29

Glass tubes for microinjection and instruments to hold the tubes were available in 1938. L. Chabry, a nineteenth-century biologist who pioneered studies of separated blastomeres before Driesch, described capillary tubes suitable for experiments under the microscope (Chabry, 1887). McClendon, whose isolated-blastomere experiments were referred to earlier, developed a micromanipulator that permitted microsurgery on eggs firmly held by the apparatus (McClendon, 1907, 1908).

Marshall A. Barber described a microinjection apparatus that permitted an experienced operator to enucleate a cell and transfer material from one cell into another (Barber, 1911). The Barber equipment was known to American cell biologist Robert Chambers. Chambers published several studies during the second decade of this century that, among other things, described the microdissection of living cells and the "sucking" of a nucleus into a capillary tube of fine bore. The nucleus was disposed of by "blowing it out" (Chambers and Chambers, 1961). Then a new machine was devised for the rapid production of microneedles and microcapillary tubes. The machine, developed by Delafield DuBois of Washington Square College, was adopted for cell studies by Chambers.

Chabry, McClendon, DuBois, and Barber were not alone. J. Comandon and Pierre de Fonbrune of the Pasteur Institute, Paris, described the manufacture of micropipettes, microneedles, and other microscopic glass tools during the 1930s (de Fonbrune, 1949). Not only were they engaged in tool making, but they were enucleating and transplanting nuclei in amoebae (Figure 2-5). More recent ac-

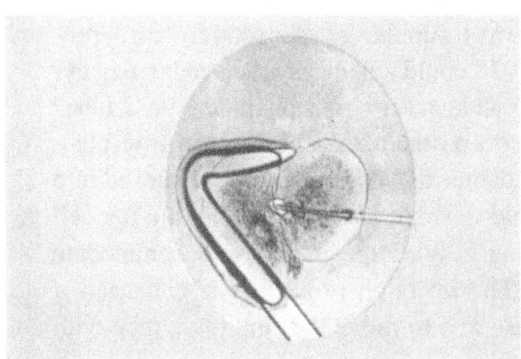

FIGURE 2-5. *Nuclear transplantation in an amoeba. The recipient amoeba (left) is held in position with a glass hook. The nucleus of the donor amoeba (right) is pushed gently into the recipient cytoplasm with a glass needle. (From de Fonbrune, 1949)*

counts of nuclear transplantation studies of amoebae can be found in Yudin (1979) and Muggleton-Harris (1979). Because of the relatively long history of experiments with cell microsurgery, it seems curious that Spemann, distinguished Nobel laureate, seemed unaware of how to get a nucleus into an enucleated egg.

Vegetative propagation, a form of plant cloning, is old. Animal biology frequently lags behind botany. However, by the late nineteenth century, the question that led to animal cloning was posed by Weismann and others, and preliminary answers were provided by blastomere separation. Better answers were forthcoming with observations of herniated eggs and constricted zygotes. It was but a brief pause before Briggs and King reported from Philadelphia their first successful nuclear-transplantation experiments (Briggs and King, 1952).

Should we have taken the first step? Does the question refer to human manipulation of plants for over four millennia? Does it refer to separated blastomeres? Or the primitive nuclear-transplantation experiment of a constricted zygote? Or the pre–World War II nuclear-transplantation experiments with amoebae in Paris? These early steps comprise an important portion of the history of experimental embryology. The experiments are important because they are the foundation of modern reproductive biology.

3 To Clone a Frog

The historical account of the experiments that led to frog nuclear transplantation explains why cloning was attempted. It does not reveal *how* cloning was accomplished. An account of the nuclear-transplantation procedure, "the how of cloning," provides insight into the purposes, opportunities, and limitations of the cloning technique.

Nuclear transplantation refers to the process of moving a nucleus from one cell to another. The transfer of a nucleus would have little biological meaning if it were not from one *kind* of cytoplasm to another *kind* of cytoplasm. Why was *egg* cytoplasm chosen as the recipient cytoplasm for inserted nuclei? Why were *blastula* nuclei studied first?

EGG CYTOPLASM

If one can enucleate an amoeba and place into that amoeba the nucleus from another amoeba, then there is little, if any, practical reason why an embryonic nucleus could not be inserted into the cytoplasm of a specialized adult cell. However, pioneer investigators chose to place an embryonic nucleus into enucleated egg cytoplasm. Their choice of egg cytoplasm was related to the fundamental question posed by these investigators. Do irrevocable changes occur in the genetic material of cells as they develop? Only if the transplanted nucleus is placed in egg cytoplasm can this question be answered.

A frog zygote nucleus in frog egg cytoplasm results in a developmental system that forms a frog. A human zygote nucleus in human egg cytoplasm results in a developmental system with the potentialities of forming a human. Egg cytoplasm has the support systems to permit programming of the zygote nucleus that results in an intact whole individual, be it frog or human. Consider, if you will, transplanting an embryonic frog nucleus into the cytoplasm of an enucleated brain cell. It requires little imagination to realize that brain-cell cytoplasm probably does not have an adequate supply of stored energy to allow for the development of a frog. Probably more crucial is the possibility that the cytoplasm emits signals to the nucleus that *preclude* expression of many genetic characteristics. This possibility seems particularly compelling as more is learned about cytoplasmic control of genetic activity.

Egg cytoplasm would be an inappropriate test site for the study of the capacity of a nucleus to guide development if the cytoplasm were preprogrammed for substantial development. Therefore, it is important to ask how much can egg cytoplasm develop without any nucleus at all.

Unfertilized eggs of many species develop rather well. As noted, Jacques Loeb obtained swimming larvae from unfertilized sea urchin eggs treated with seawater and magnesium chloride, proving that fertilization is not necessary for early development in this species. The chromosomes of the egg can sustain substantial development. There are natural populations of lizards which survive with no males at all (Cole, 1984). Thus, parthenogenesis is found in nature as well as in the laboratory.

Will eggs deprived of sperm and bereft of their own chromosomes develop? J. F. McClendon provided some early observations on the behavior of eggs with no hereditary material. McClendon removed the maternal hereditary material from starfish eggs. The enucleated eggs were then *stimulated* with carbonated seawater. Development was initiated and a small cluster of non-nucleated cells was formed (Figure 3-1) (McClendon, 1908).

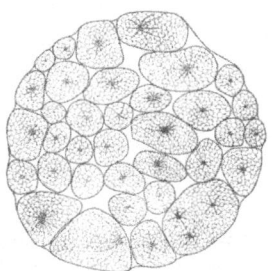

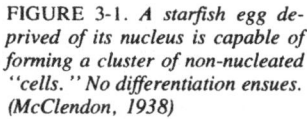

FIGURE 3-1. *A starfish egg deprived of its nucleus is capable of forming a cluster of non-nucleated "cells." No differentiation ensues. (McClendon, 1938)*

33

Ethyl Browne Harvey, an embryologist at Princeton University, continued and extended the studies of eggs that develop without genetic material. Not unlike McClendon, Harvey treated non-nucleated fragments of sea urchin eggs with chemical agents that *stimulate* development. A sea urchin egg, lacking *both* maternal and paternal hereditary material, when stimulated, will divide and form a ball of cells. No muscle, no nerve tissue, no skeleton forms. Cell division, but no cell specialization occurs in these curious non-nucleated sea urchin embryos (Harvey, 1940). Sea urchins are a less complex form of life than frogs. Will frog eggs develop if they have no nucleus?

A haploid individual is one with chromosomes from only one parent. (Figure 3-2.) Haploid frog embryos develop regularly to the

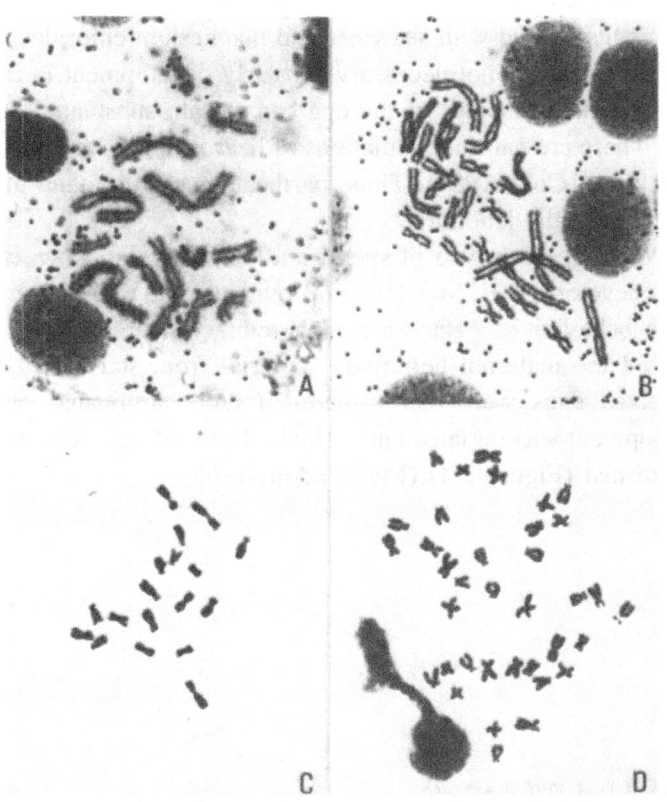

FIGURE 3-2. *Radiation ablation of maternal chromosomes results in haploid embryos of the leopard frog (A) and the South* *African clawed toad (C). Unirradiated diploid controls are shown at (B) and (D). (From McKinnell, 1981)*

swimming tadpole stage but ordinarily no further. However, it was reported in Japan that a haploid tadpole had undergone metamorphosis to become a juvenile frog (Miyada, 1960). A diploid individual contains chromosomes from both male and female parents. In the laboratory diploid development ordinarily follows when sperm are spread across freshly ovulated frog eggs. A number of years ago, American embryologist and cell biologist Keith Porter noted that maternal chromosomes approached the surface of the frog egg shortly after sperm were placed on the egg. The maternal chromosomes were thus accessible to a microneedle. Porter removed the maternal chromosomes with the microneedle, and the surgically treated egg developed as a haploid (Porter, 1939). This kind of haploid development is known as androgenesis, meaning that only the *paternal* chromosomes participate in development. It is, then, relatively simple to produce a tadpole that develops with only one set of chromosomes. However, the question posed was will a frog egg develop with no nucleus, with no chromosomes whatsoever?

The answer came in an interesting study by Robert Briggs, Elizabeth Ufford Green, and Thomas J. King (1951). These investigators treated sperm with toluidine blue, a dye that has an affinity for nuclei. It will stain the nucleus of a sperm, inactivating the hereditary material of the sperm without doing much damage to the sperm's motility or its capacity to penetrate an egg. What happens when an egg, "fertilized" with a sperm that contributed no chromosomes, is enucleated according to Porter's procedure? The egg, "fertilized" with sperm containing no functional *paternal* chromosomes, is then surgically treated for removal of the *maternal* chromosomes. The egg cleaves. It forms two blastomeres, four blastomeres, eight blastomeres—it forms a ball of cells, and the final stage is an imperfect blastula. Since tissue differentiation begins *after* the formation of a blastula, nucleusless frog eggs divide but they do not differentiate. These results with nucleusless frog embryos are thus essentially identical with the results obtained by McClendon and Harvey with nucleusless starfish and sea urchin embryos.

We now have our answer to the question of why egg cytoplasm was (and is) used. Egg cytoplasm deprived of its own chromosomes seems to be an ideal environment in which to test the differentiative capabilities of an embryonic or adult nucleus because, lacking chromosomes, it does not have the capacity to differentiate. However, egg cytoplasm *with* competent chromosomes forms a harmonious developmental system that has the capacity, in the proper environment, to develop into an adult organism.

BLASTULA NUCLEI

Why did early nuclear transplanters clone nuclei derived from blastulae? Why didn't they transplant nuclei from older embryonic stages? There are two good reasons: one is that blastula cells are considered to be undifferentiated. Spemann, who did the primitive nuclear-transplantation experiments with constricted zygotes, also exchanged patches of cells between young embryos. Even at the early gastrula stage (a stage in development that follows the blastula), cells destined to form brain if left undisturbed will form skin when transplanted to skin-forming areas, and cells destined to form skin when left in place will form brain when surgically inserted in that portion of the embryo where brain is formed. Results are different when similar exchanges of tissue are made between older embryos. Late gastrula cells destined to form brain become brain even when transplanted to an anatomically inappropriate site. These observations, as well as many others, led embryologists to believe that blastula cells are not yet differentiated, have not yet developed specialized cell function. Thus these cells are appropriate donors of nuclei when one attempts to see if the nuclear-transplantation system works.

If older, differentiated nuclei were initially chosen for cloning, and if development of the clone was imperfect, investigators wouldn't know if the imperfections of the experimental embryo resulted from the kind of nucleus that was transplanted or from the transplantation procedure itself.

The second reason for studying blastula nuclei is that they have already undergone twelve or thirteen cell generations. This may not sound impressive, but these generations produce a blastula composed of several thousand cells (calculate 2^{12} or 2^{13}). The ability of investigators to characterize the genetic determinants of a blastula nucleus was an enormous stride forward beyond Spemann's 16- or 32-cell stage constricted egg experiment. Substantial new information became available when the nuclear transplantation of blastula nuclei was successful.

VALIDITY OF BLASTULA NUCLEAR TRANSFERS

Robert Briggs and Thomas J. King reported to the scientific community in 1952 that normal tadpoles developed from their nuclear-transplantation experiments. How is the validity of such an experiment determined?

A valid cloning experiment produces a frog asexually. Since cloning experiments seek to test the developmental capacity of the transplanted nucleus, neither a sperm nor an egg nucleus may participate in the development of a frog that is truly a clone. There must be convincing evidence that sex-cell nuclei did not participate in the formation of nuclear-transplant frogs. The experiments described in the following discussion provide compelling evidence that cloned frogs are indeed cloned.

Nuclear transplantation requires that the recipient egg be enucleated. Aristotle knew that bee drones were produced by queen bees without copulation. Loeb produced fatherless sea urchins by treating unfertilized eggs with seawater and magnesium chloride. Tadpoles of frogs result after parthenogenetic stimulation of unfertilized eggs. There are reptiles that reproduce without males (Moritz, 1983; Cole, 1984) and even mice have been produced experimentally by parthenogenesis (Chapter 5). Hence, extraordinary care was taken to ensure that maternal chromosomes were eliminated from the recipient, unfertilized egg, to exclude the possibility of parthenogenesis. Scientists wanted to determine what the developmental potential of the *inserted nucleus* was. They already knew the capacity of maternal chromosomes functioning alone.

As noted earlier, the production of androgenetic haploids involves removing the *maternal* chromosomes after fertilization. The skill of haploid production witnesses to the ability of experimenters to recognize and remove maternal chromosomes. (Figure 3-2). Briggs and King reported that *all* the surviving tadpoles, in which they attempted to remove the maternal chromosomes from *fertilized* eggs, were haploid. Thus they indeed demonstrated their ability to obliterate the maternal hereditary apparatus. In their cloning experiments, in contrast to the haploid enucleation manipulations, the recipient eggs *were not fertilized*. Hence, the eggs could be considered nucleusless before the nuclear transplantation procedure.

If development that occurs after nuclear insertion is attributable to the transplanted nucleus (and not to an inadvertently retained set of maternal chromosomes), then it should be constrained to follow the developmental capacity of the donor nucleus and, as such, is nuclear specific. Hybrids have proved useful in establishing the specificity of nuclear development.

Hybridization has been studied extensively in amphibia (Moore, 1955). Since eggs and sperm of frogs and toads are readily available to experimenters, it is relatively simple to produce a hybrid by combining sperm from one species with eggs from another. Some combinations result in viable hybrids, others in lethal hybrids. Certain lethal hybrids are interesting to biologists because they develop to a particular embryonic stage and then development is arrested. The experiment that follows utilized the arrested development of a hybrid.

I have said that if development of the recipient egg is a function of the nucleus inserted into the egg, then development is limited by the kind of nucleus that is inserted. The hybrid formed with the sperm of a bullfrog (*Rana catesbieana*) and the egg of a leopard frog (*Rana pipiens*) arrests at the end of blastulation or the onset of gastrulation. As noted earlier, many profound cellular and biochemical changes occur in the gastrula stage. What happens when the nucleus of the hybrid (bullfrog sperm crossed with leopard frog eggs) blastula is injected into an enucleated egg of a leopard frog? The cloned embryo invariably arrests at the onset of gastrulation (King and Briggs, 1953). Control nuclear transfers, that is blastula nuclei from non-hybrid donors inserted into the appropriate egg cytoplasm, develop normally. Thus the nuclear-specific pattern of development is substantial genetic evidence that growth of the operated eggs was truly attributable to the inserted nucleus.

There is a way of permitting normal development of the cloned embryo *and* having direct genetic evidence that development of that embryo is due to the hereditary material inserted. In Minnesota and some contiguous states, there are pigment mutants of the common leopard frog and, of course, the ordinary spotted variety (Figure 3-3). These mutants are known as Kandiyohi (Figure 3-4) and Burnsi (Figure 3-5) (McKinnell and Dapkus, 1973). Nuclei from the mutants have been used as genetic markers in cloning experiments.

The hereditary material of ordinary leopard frogs dictates that

FIGURE 3-3. *The northern leopard frog,* Rana pipiens, *with large and distinct spots on its body and legs. This is the species that was used in the first successful cloning experiments.*

FIGURE 3-4. *The mottled Kandiyohi mutant of the northern leopard frog,* Rana pipiens, *that occurs principally in Minnesota. (Photograph by Gordon A. F. Dunn)*

FIGURE 3-5. *Northern leopard frogs with reduced or no spotting are known as Burnsi mutants and occur principally in Minnesota. (Photograph by Gordon A. F. Dunn)*

39

FIGURE 3-6. *A cloned frog produced from an enucleated egg of a Vermont spotted leopard frog and an inserted nucleus of a Minnesota Kandiyohi mutant embryo donor. The expression of the mutant characteristic (Kandiyohi) is genetic evidence of the authenticity of the cloning procedure.*

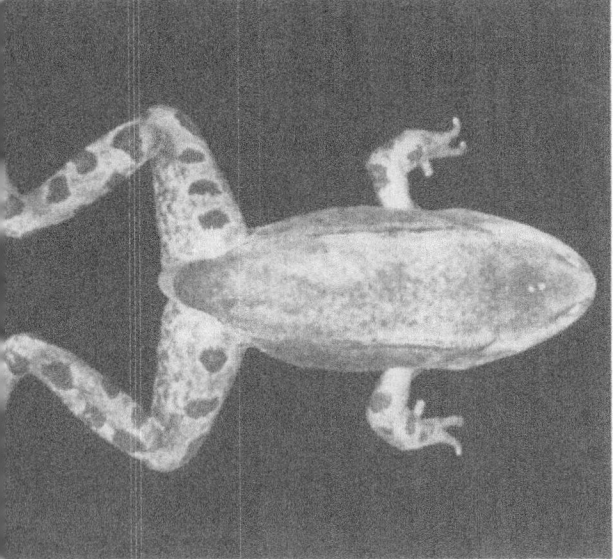

FIGURE 3-7. *Nuclear transplant frog with Burnsi phenotype. The frog was produced by inserting an embryonic nucleus obtained from the mating of adult Burnsi frogs into an enucleated egg of a wild-type female. The dorsal surface of the frog is free of spots. (From Simpson & McKinnell, 1964)*

the developing frog will have leopardlike spots. Another way of stating this fact is to say that expression of the spotting patterns of frogs depends upon nuclear function. This fact is no less true of mutant variants than it is of ordinary spotted frogs. Hence it is possible to use the activity of the mutant gene as a "marker" for a transplanted nucleus.

If nuclei from Kandiyohi or Burnsi mutant blastulae are inserted into enucleated eggs of ordinary frogs, the development of the resulting nuclear-transplant frog is determined by the nuclear donor, not by the genome of the female that provided the recipient egg. Thus the nuclear-transplant frogs will have the pigment pattern of either the mottled Kandiyohi nuclear donor (McKinnell, 1960) (Figure 3-6) or the near-spotless Burnsi nuclear donor (Figure 3-7) (Simpson and McKinnell, 1964) not the pigment pattern of an ordinary leopard frog. Experiments with Minnesota mutant frogs offer compelling genetic proof that the inserted nucleus provides the genetic instructions for the growth of the frog.

Other pioneer nuclear transplanters used different genetic tags or markers associated with the donor nucleus to provide persuasive evidence of the validity of cloning. For example, nuclei of most cells of the South African clawed toad, *Xenopus laevis*, contain two minute structures known as nucleoli. There is a mutant that has cells which usually have only one nucleolus (Elsdale et al., 1958). When the nucleus of a one-nucleolus mutant is inserted into an enucleated egg of a two-nucleoli female *Xenopus* and the resulting tadpole has only one nucleolus per cell, then we are justified in believing that it was the inserted nucleus, not an inadvertently retained maternal egg nucleus, that determined the development of the tadpole (Gurdon et al., 1984). Cells with extra sets of chromosomes, obtained from polyploid donors, have also been used as nuclear markers in frogs and salamanders (McKinnell et al., 1969a; McKinnell and Tweedell, 1970; DiBerardino et al., 1983). In all experiments with nuclear markers, the conclusion that follows careful observation of experimental embryos and frogs has been that it is the donor nucleus that guides or programs development. Because it is a body-cell nucleus, not a gamete nucleus, the cloned embryo is asexually produced.

The enucleation procedure is valid, and there is commanding evidence of several kinds that provides assurance that the inserted nucleus guides development. So, if the cloning procedure works for the American leopard frog, can it work with other species? In the years following the prototypic Philadelphia cloning experiments, the technique was extended to many different amphibian species. Already mentioned was the South African clawed toad, *Xenopus laevis* (Elsdale et al., 1960). To this may be added various species of frogs and toads from America, Europe, and Asia, as well as of salamanders (see Appendix IV, McKinnell, 1978). It may be interesting to note that nuclear transfer has also been studied in the fruit fly, *Drosophila* (Illmensee, 1973). Cloning studies in fish are underway in the U.S.S.R. (Gasaryan et al., 1979) and the People's Republic of China (Research Group of Cytogenetics, Academy Sinica, 1980), and cloned mice have recently been reported (Illmensee and Hoppe, 1981; McGrath and Solter, 1983a & b) (Chapter 5). However, thus far, frogs, toads, and salamanders remain the most studied cloned animals.

All the genetic manipulations discussed in this section demonstrate that the development observed in cloned animals is authentically attributable to the inserted nucleus. How, then, is cloning done?

THE CLONING PROCEDURE

How difficult is the cloning procedure in frogs, toads, and salamanders? It is very difficult, but simpler than mouse cloning. This fact has certain implications for the plausibility of claimed human cloning—which will be discussed later. Cloning amphibia was successful over 30 years ago, but mammalian cloning remains in a primordial state. The success with frogs before mice may be attributed to a century of experimentation in the reproductive biology of amphibians. Before the first successful frog-cloning experiments were performed, a number of procedures had been developed that would become useful to the craft of nuclear transplantation. These included the ability to obtain eggs and sperm from frogs, *in vitro* fertilization, removal of maternal chromosomes from eggs, and dissociation of embryos into individual cells. Further, amphibian eggs and cells are substantially larger than mouse eggs and cells—large eggs and large cells are easier to manipulate than small eggs and small cells. Finally, frog embryos develop in water and are easy to observe. Mouse and human embryos ordinarily develop in the uterus and are usually hidden from direct observation. Thus, it is not surprising that frogs were cloned first.

Obtaining Frog Eggs for Cloning Experiments

The northern leopard frog, *Rana pipiens*, spawns in the spring, and no eggs can be ovulated immediately after spawning. The frog spends the summer months foraging in the fields and growing eggs. By the time fall arrives, the eggs have grown to their maximum size, and the frogs are ready for hibernation under the ice of lakes and streams. Although egg release will not occur spontaneously until the next spring, ovulation can be induced from September to, or past, the time of natural ovulation. Eggs leave the ovary, move to the reproductive tubes, and become available to the embryologist when the female frog is injected with pituitary glands or a combination of pituitary glands and the hormone progesterone. The eggs can be extruded from the female after this treatment by gently squeezing the abdomen.

Obtaining Frog Sperm

Sperm can be obtained by cutting frog testes into fine pieces in a diluted salt solution. The testes (testes is plural for testis, the male sex-cell gland; testicle is a diminutive and not a particularly appropriate word for a sexually mature beast) are dissected from the male, which usually requires sacrifice of the frog donor. Sperm may be obtained without sacrifice by hormonal release. This procedure is used in my laboratory to spare the increasingly scarce frogs found in nature. Chorionic gonadotropin (a commercially available hormone present in pregnant humans) is injected into a mature male frog. Within one hour, motile sperm are released from the testes of the frog and are found in the urine (McKinnell et al., 1976a). The sperm are capable of fertilizing frog eggs, and the experiment, in addition to being useful to the cloner, shows that reproductive glands of evolutionarily primitive cold-blooded frogs respond to hormones of evolutionarily advanced warm-blooded human females (Roth et al., 1982). This attests to the unity of life and the similarity of biologic processes among the vertebrates since ancient times.

Fertilization and Husbandry of Frog Embryos

Eggs and sperm can be combined in a glass dish at a carefully predetermined time. By caring for the fertilized eggs at a particular temperature (usually 18°C) and a particular time, donor embryos of predetermined stage of development can be obtained.

There is no need to provide an elaborate culture medium for developing frog embryos, as one must do with the much smaller mammalian embryos. Simple glass dishes and water are adequate because frog eggs are large and contain stored food enough for about

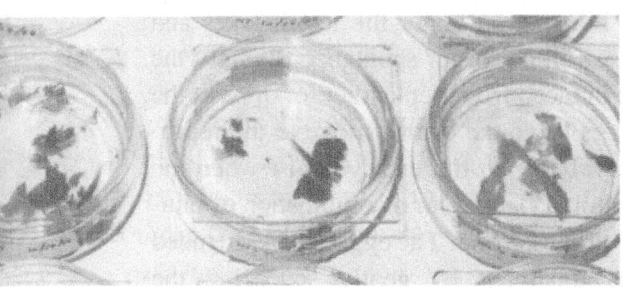

FIGURE 3-8. *Tadpoles are herbivores. They will eat, among other things, cooked lettuce.*

12 days. Frog tadpoles are vegetarians when they begin feeding and survive nicely on cooked lettuce or spinach (Figure 3-7) until they become carnivores at metamorphosis about 90 to 100 days after fertilization.

Preparing Eggs to Receive a Transplanted Nucleus

Freshly ovulated eggs, contrary to what many textbooks state, have the same amout of DNA as an ordinary body cell. That amount of DNA is twice the amount contained in a sperm; hence, it is called diploid. A sperm has the haploid amount of DNA. Diploidy in freshly ovulated eggs is a boon to the experimenter. If diploid eggs combined with diploid sperm, the amount of DNA of the resulting individuals would become enormous in only a few generations. This, of course, does not occur. What happens is that the final maturation of the frog egg, to become haploid as the sperm already is, occurs *after* it is released from the ovary and at that time it is activated by the penetration of the sperm.

Embryologists can *mimic* the activation of sperm penetration by pricking the surface of the freshly ovulated egg with an extremely fine glass needle. The maternal chromosomes respond to the needle prick as though it were a sperm penetration, by approaching the surface of the egg in what would ordinarily be an attempt to extrude half the chromosomes, thus half the DNA, in a minute cell known as a polar body. It is at this point that the maternal chromosomes and the chromosomal DNA of the unfertilized egg are accessible to either surgical removal or removal by the flash of a laser apparatus mounted on a microscope (Figure 3-9). Removal of maternal chromosomes and chromosomal DNA, whether by surgery or laser (McKinnell et al., 1969b; Ellinger et al., 1975) results in an egg devoid

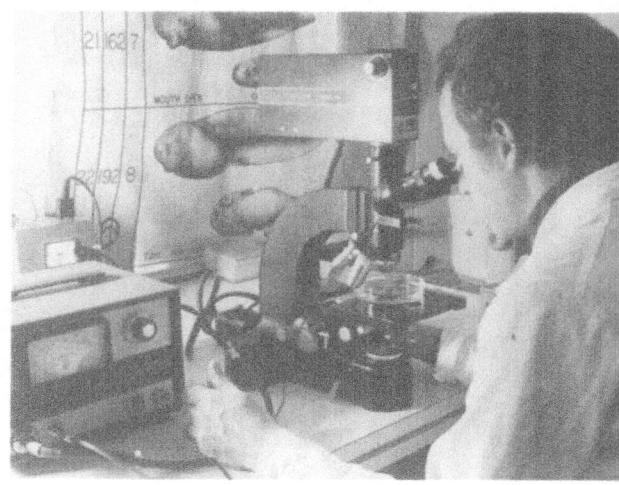

FIGURE 3-9. *Enucleation of an unfertilized egg can be accomplished by microsurgery or by irradiation with a ruby laser apparatus mounted on a microscope.*

of any genetic material in the form of chromosomes. The enucleated egg now needs only to be removed from its jelly envelope (by cutting with very fine scissors and forceps) to be a suitable recipient for a transplanted nucleus.

Preparing the Donor Nucleus for Transplantation

The eggs that were fertilized are permitted to attain an appropriate embryonic stage to serve as nuclear donors. This need not take long. An embryo composed of several thousand cells develops in less than 24 hours at 18°C.

When a donor embryo is selected, the jelly coat is removed and the appropriate region of the embryo is dissected and placed in a solution that causes the dissociation or separation of individual cells. A physiological salt solution containing neither calcium nor magnesium salts is used. Young embryos in a calcium- and magnesium-free solution rapidly dissociate to individual free cells. With older embryos, it may be necessary to add special substances (such as agents that chemically bind free calcium in the solution or protein-digesting enzymes that break down the complex material that binds cells together into a tissue fabric) to the dissociation solution to obtain free cells.

It is relatively simple, in principle, to bring together the separated donor cells that provide nuclei and the enucleated host eggs.

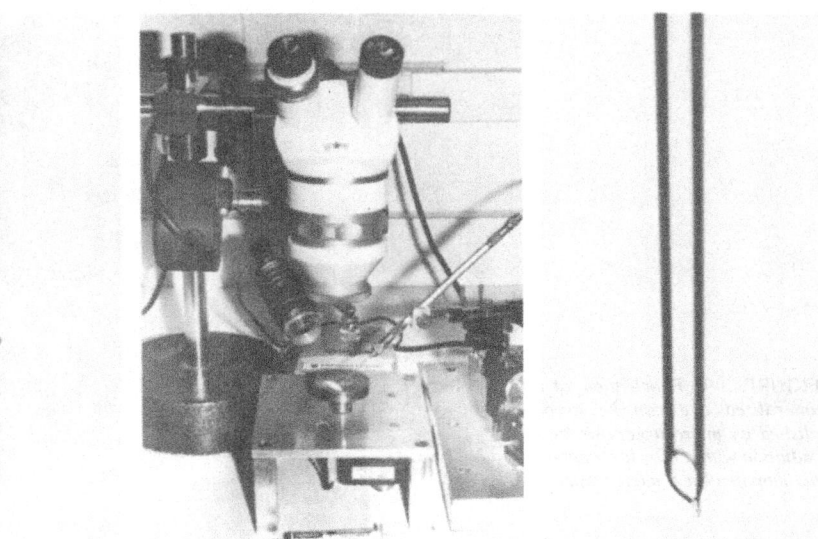

FIGURE 3-10. *A 37 micrometer pipette with tip beveled by grinding with diamond paste on a revolving turntable.*
FIGURE 3-11. *Revolving turntable used with diamond paste for grinding beveled tips on micropipettes. The micropipette is* held in position with a micromanipulator, and the progress of grinding is observed with the aid of the dissecting microscope.
FIGURE 3-12. *Beveled micropipette with a tip sharpened with a microforge.*

The mechanics of the conjugal surgery involve micropipettes, micro-injection apparatus, and micromanipulation equipment.

A pipette is a small glass tube used to transfer liquids. A nuclear-transplantation micropipette differs from an ordinary pipette by its size (it is tinier) and by the acuteness of its tip (it is very sharp). The production of a nuclear-transplantation micropipette involves heating glass tubing to a temperature just less than melting so that the softened tubing can be pulled to a small diameter. A diameter suitable for blastula nuclei is 37 micrometers, a diameter a bit larger than 1/1000 of an inch. The blunt tip of the tiny glass tubing is ground to a bevel (Figure 3-10) on a turntable with diamond paste (Figure 3-11). The beveled micropipette is sharpened (Figure 3-12) by touching its tip to a heated filament in an apparatus called a microforge. The diameter of the cell chosen as a nuclear donor will be somewhat larger than the diameter of the orifice of the micropipette. The micropipette is positioned adjacent to the selected dissociated cell with the microinjection apparatus.

The microinjection apparatus (Figure 3-13) is simply a machine, a micromanipulator, that holds a tool (in this case a micropipette) very steady, and permits small and precise movements of that tool. The donor cell is drawn into the micropipette with the microinjection apparatus. The procedure is like squeezing the rubber bulb of

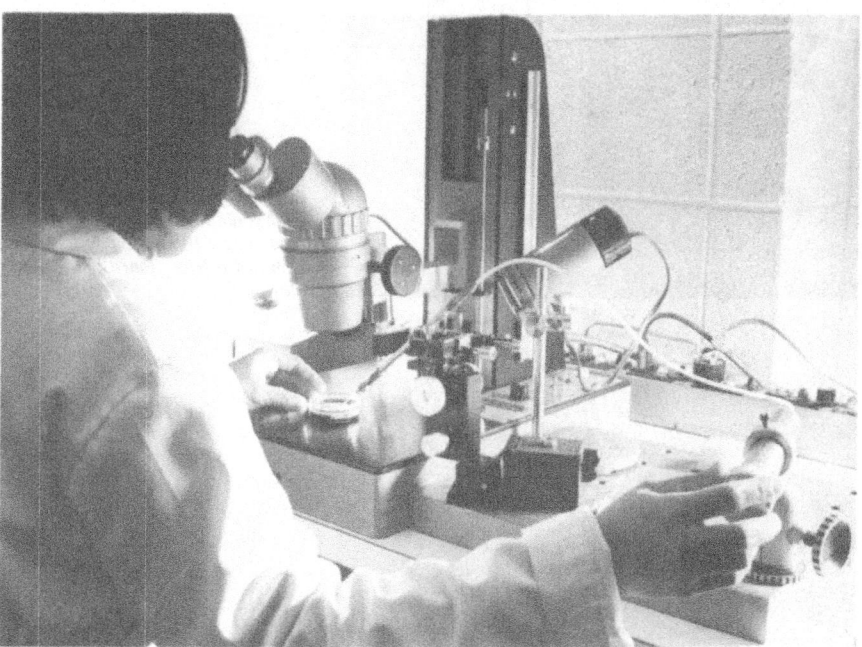

FIGURE 3-13. *Nuclear insertion is accomplished with the aid of a microinjection apparatus and microscope. The scientist has her hand on a heavy-duty syringe which is connected to a micropipette held in position with a micromanipulator. The electrically focused microscope is foot-operated, which permits the scientist to use both hands in the cloning procedure.*

a medicine dropper in order to draw fluid into the dropper. When the cell enters the orifice of the micropipette, the cell membrane is ruptured (recall that a cell larger than the micropipette orifice was chosen), and there is some dispersal of the donor-cell cytoplasm. The cell membrane is very thin but very important. If it is left intact on the inserted donor cell, the enucleated ovum with its donor cell cannot develop. However, a donor cell with its nucleus liberated by virtue of a broken cell membrane can interact with the egg cytoplasm to form a viable developmental system—sometimes resulting in the

FIGURE 3-14. *A frog results from the successful insertion of an embryonic nucleus into an enucleated egg. (Photograph by Kris Kohn of a frog cloned by Cynthia Horton.)*

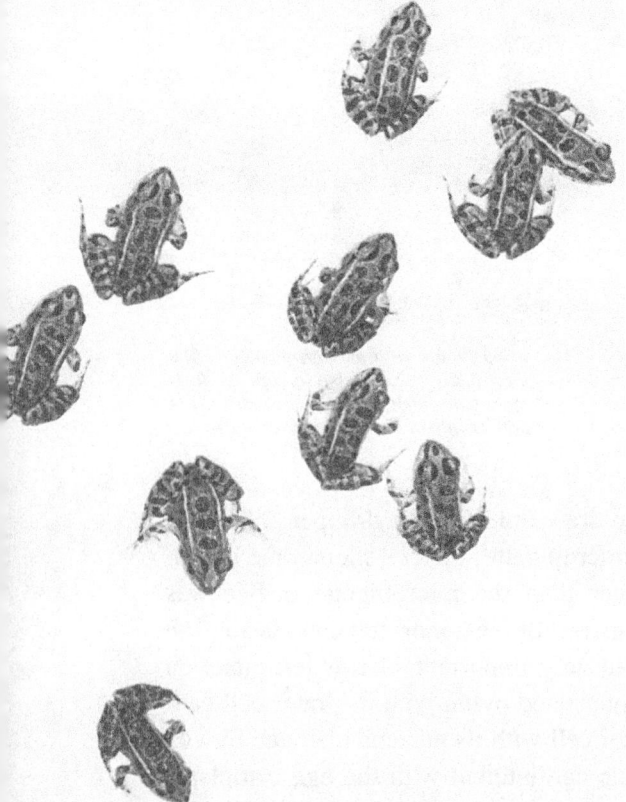

FIGURE 3-15. *Several frogs develop from nuclei of a common blastula. Although the frogs are isogenic (i.e., genetically identical), they vary with respect to spot distribution on their dorsal body surface. (Photograph by Gordon A. F. Dunn)*

formation of a frog (Figure 3-14). More than one frog ensues, each genetically identical to the others, when several nuclei from a single embryo donor are separately inserted into enucleated ova (Figure 3-15). The set is collectively known as a clone.

The Skill Required for Nuclear Insertion

Cells are microscopic. Nuclei are microscopic. The maternal chromosomes are microscopic. The movements of the micropipette are microscopic. Thus the procedures just described must be performed under the relatively high magnification of a light microscope. Although cloning is conceptually simple, the operation is exquisitely complex; and much skill, coupled with precision, is required of the operator.

Cloners not only have to be expert at reproductive biology and cell surgery, they must also be proficient enough to make the implements of the craft. Probably the most difficult implement to make is a beveled and sharpened micropipette with an orifice appropriate for an embryonic cell. Consider the skill required to make an ordinary medicine dropper. Add to that skill the ability to change the orifice of the dropper to a sharpened bevel. Now consider the skill required to accomplish all of this but also to make the beveled and sharpened dropper so tiny that, when properly used, it can break a single cell while not damaging the nucleus contained within. That is why micropipette-making is exacting.

It is extraordinarily difficult to penetrate the surface of an enucleated egg with an improperly sharpened and beveled micropipette. The otherwise delicate egg has a membrane that is tough and very stretchable. A dull micropipette depresses the surface of the egg with pressure by the micromanipulator. The depression becomes so great that the improperly sharpened micropipette sometimes goes all the way through the egg and protrudes from the other side without having penetrated the egg membrane. Or the pressure exerted from a dull micropipette may rupture the egg membrane, and egg cytoplasm spews into the operating dish. These are certainly unacceptable results and ordinarily do not occur with a finely sharpened micropipette. The procedure is diagrammed in Figure 3-16.

After the nucleus is transplanted, there is little to do other than record growth and development. Frog husbandry, like the care of

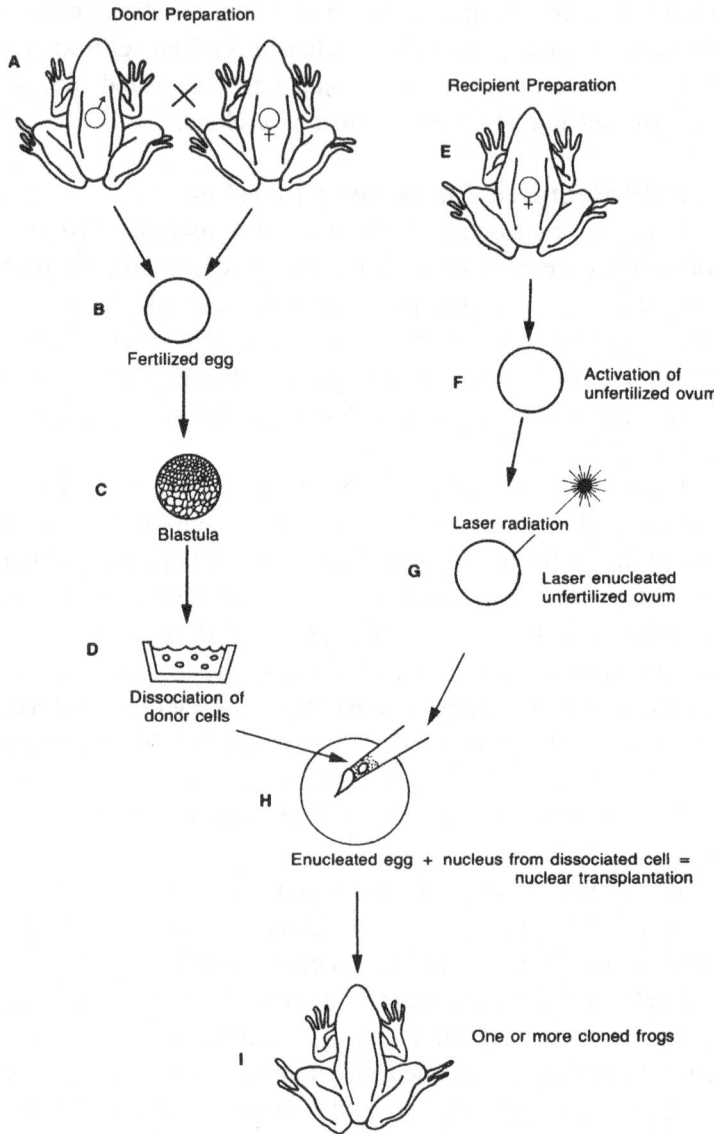

FIGURE 3-16. *Nuclear transplantation in the leopard frog. Adult frogs (A) are mated in the laboratory and fertilized eggs (B) are permitted to develop to the blastula (C) stage. The blastula is dissociated (D) to provide donor nuclei for transplantation. A recipient ovum is obtained from a gravid female (E), the ovum is activated by a prick of a sharp glass needle (F), and* *enucleated by the pulse of a ruby laser, or by microsurgery (G). Transplantation is accomplished by inserting the donor nucleus into the previously prepared recipient ovum (H) resulting in the production of a frog (I). Several nuclei from the blastula can be transplanted to recipient ova resulting in multiple genetically identical cloned frogs.*

mice or other laboratory animals, is expensive. It takes more than a year for a frog embryo to attain sexual maturity. However, despite the time and expense involved in rearing, a few cloned frogs have grown to sexual maturity in the laboratory. (Figure 3-17)

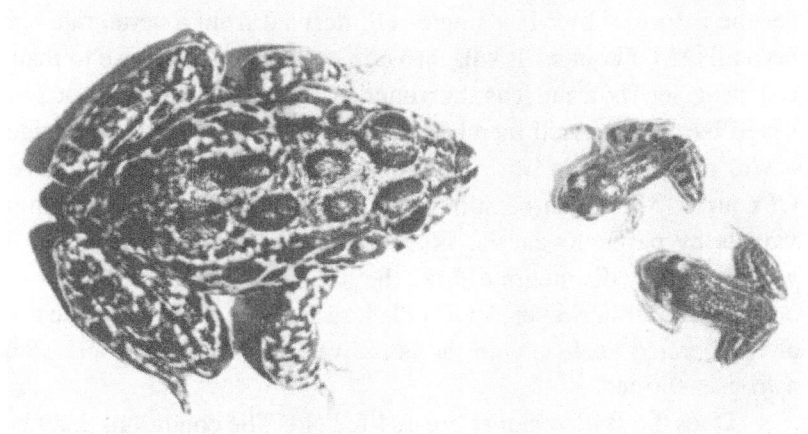

FIGURE 3-17. *Sexually mature frog produced by cloning. The small frogs are two* *of many offspring fathered by the clone. (From McKinnell, 1962)*

This brief account of frog cloning is not intended to be a "how to" manual. The procedure is much more difficult than the description makes it seem. It requires great skill and much practice. Few investigators succeed without instruction from experienced cloners. Those investigators who seek more detailed technical information on frog nuclear transplantation are referred to my monograph on the subject (McKinnell, 1978).

SIGNIFICANCE OF TRANSPLANTS

What do the frogs formed from cloned blastula nuclei show? The frogs witness to the capacity of skilled embryologists to take apart an early embryo to the level of whole but separated cells, to break one of those cells in order to free its nucleus, to place that liberated nucleus into a previously prepared enucleated egg, and to effect the transplantation with such dexterity that the result is a viable embryo with the capacity to develop into an adult. More important than the skill of the embryologist is the biological significance of the

51

cloned frog. The frog attests to the completeness of the genetic material of the blastula nucleus from which it was derived. This demonstrated intactness of blastula cell DNA is in harmony with the early experiments of Driesch, McClendon, Loeb, and Spemann on younger developmental stages.

From time to time, biologists, like most thoughtful people, ponder the nature of life. Is a single cell, derived from a vertebrate embryo alive? Of course. It will thrive in culture and give rise to many cell progeny. Is a nucleus, surrounded by cytoplasm, but not protected by an intact cell membrane, alive? Hardly. It will not divide. It will disintegrate in just a few hours. Is an unfertilized egg alive? Of course. Many can be stimulated to divide and give rise to individuals by parthenogenesis. Is an *enucleated* egg alive? Hardly. It will eventually disintegrate. Like the liberated nucleus, it has no independent life. However, let a skilled microsurgeon combine the not-alive liberated nucleus with the not-alive enucleated egg—and often a frog is formed.

Does the skilled cloner create life? No. The conditions that permit life are *restored* by cell surgery—and I suppose that is one reason why cloning is fascinating to experimenters. They seek answers to significant questions, and each experiment is a test of their ability to restore conditions that permit a spark of life to flame as a living substance capable of developing into a mature organism.

Cancer, Aging, and Other Challenges

The success of nuclear transplantation with frog blastula cells encouraged researchers to do similar experiments with older, more specialized nuclei. Adults are composed of many kinds of specialized (differentiated) cells. If, as some believe, certain kinds of cancer result from the improper functioning of the differentiation process, then perhaps some cancer can be viewed as tissue specialized in ways that are harmful to humans. Humans age, and the changes that occur in the cells of older people may be viewed as another kind of non-beneficial specialization. The cloning procedure has the potential of revealing to what extent cell specialization is attributable to alterations in the genetic material. Therefore, a variety of specialized cells, including cancer cells, have been studied by cloning. A cloning study relating to aging cells has begun in my laboratory at the University of Minnesota.

The cloning procedure has been used productively in several areas of research: differentiation—much has been learned in an area that stimulated the early cloning experiments; cancer—a much studied problem; immunobiology—only tentative, but promising probes have been made thus far; and aging—work which is exciting to me and still in its early stages.

Before discussing cloning experiments, let me point out that studies in genetics and cell biology can be conducted on several levels of complexity. An ordinary microscope with glass lenses and visible light has a maximum useful magnification of about 1000. It is possible, for example, using fixed and stained slides, to determine with an ordinary microscope whether nuclear DNA is fairly evenly distributed or is clumped. For fine anatomical detail, an electron microscope is needed. For example, the question of whether viruses are present in nuclei prepared for examination must be answered by the electron microscope, not the light microscope, because viruses are too tiny to be visualized with an ordinary microscope. A molecular biologist with centrifuge, electrophoresis apparatus, and other paraphernalia may wish to describe patterns of DNA replication in different kinds of cells and to gain insight into the nature of cell function by analysis of the replication patterns. I mentioned this not exhaustive inventory of cell-probe procedures to make a point. The fate of a cell and its nucleus is fixed when one prepares it for examination by conventional microscope or molecular procedures. Elegant as is the image obtained with the electron microscope, it can suggest only what the nucleus *was*. The molecular biologist analyzes an extract from cells and trusts that the findings have some relationship to the previously living cells.

The cloner, in contrast to some other cell biologists, asks what a nuclear transplant egg *will become*, not what it has been. The system is alive and the answers are provided by the living embryo that results from cell surgery. This is not to say that *all* cell biologists work with lifeless material. Tissue-culture studies of cell populations suspended in flasks of nutrient material, or growing as monolayers in flat dishes containing appropriate liquid cuisine, certainly relate to life. Details of the anatomy of these living cells in culture can be observed with a special microscope, the phase-contrast instrument, that does not involve killing cells in order to observe them. However, cells in culture do not form intact and functional embryos and adults. This is perhaps another reason why nuclear transplantation is fascinating to me and others. We do not surmise that a nucleus has certain capabilities, we do not conjecture that a nucleus may interact harmoniously with certain cytoplasm. We know this as we observe living, swimming, vertebrate embryos, some of which develop to the adult state.

CLONING STUDIES OF DIFFERENTIATION

More nuclear transplantations have been done with two amphibian species than with all other species combined—the North American leopard frog, *Rana pipiens*, and the South African clawed frog, *Xenopus laevis*. These frogs are as different from each other as frogs can be. Life-style: *Rana* is terrestrial in the summer, *Xenopus* is wholly aquatic; anatomy: the *Xenopus* tadpole is primitive, with soft mouth parts and two apertures for the exit of water from the gills; the *Rana* tadpole has hard mouth parts adapted for scraping, with but a single opening for water that has passed over its gills; mating: the male *Xenopus* clutches the female at the posterior portion of her body; the amorous male *Rana* embraces the female at the level of her front limbs. Ardent males are believed to reveal something of their evolutionary history in the mode of their hug, the *Xenopus* posterior clasp being primitive, *Rana*'s embrace, advanced. There are, of course, many other anatomical, physiological, and behavioral differences of greater interest to the herpetologist than to the general reader. The differences between primitive *Xenopus* and developed *Rana* are an important factor in nuclear transplantation because if the results of the cloning were not similar, this might be attributed to their substantial biological differences. However, if the results are similar, the concurrence at least suggests that the data obtained reflect reality and that there is a possibility that similar results can be obtained with other species.

Results derived from cloning increasingly older embryonic

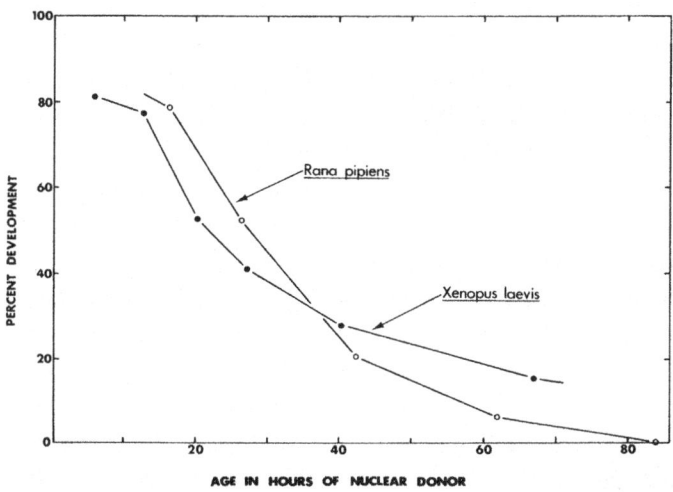

FIGURE 4-1. *The capacity of a nucleus to participate in normal development, as tested with the cloning procedure, decreases rapidly as the age of the donor embryo increases. The majority of blastulae produced by transplanting nuclear donors less than one day old develop normally. Few blastulae produced from nuclear donors two days of age or older develop normally. (From R. G. McKinnell, "Nuclear transfer in* Xenopus *and* Rana *compared." In R. Harris, P. Allin, and D. Viza, eds.,* Cell Differentiation © *1972 Munksgaard International Publisher, Inc., Denmark)*

nuclei are essentially the same in *Rana* and *Xenopus*. The success of nuclear transplantation decreases as the age of the donor nucleus increases (Figure 4-1). The yield of normally developing nuclear-transplant embryos drops precipitously as the *age of the donor nucleus increases beyond day one*. Unfortunate indeed is the fact that science-fiction writers and others who have expressed apprehensiveness about the results of cloning are unaware of this.

The facts concerning restrictions of nuclear potentialities that occur quite early in development, known since the earliest days of frog cloning (King and Briggs, 1954, 1955; Briggs and King, 1957; DiBerardino, 1980; McKinnell, 1981), have profound implications for human cloning. Some of those implications are discussed in Chapter 5.

Investigators from several laboratories have attempted to transplant nuclei from adult cells. *No normal frogs have resulted from these studies*. Some tadpoles have been produced, all of which have died. But the tadpoles are exceptionally exciting to many cell biologists. The production of tadpoles in nuclear transfer from the nuclei of diverse adult cell types such as white blood cells (DuPasquier and Wabl, 1977), red blood cells (Brun, 1978; DiBerardino, and Hoffner, 1983), skin cells (Gurdon et al., 1975), spermatogonial cells (DiBerardino and Hoffner, 1971), and kidney cancer cells (McKinnell, 1979; DiBerardino et. al., 1983) reveals that nuclear activities can be reprogrammed, at least to a limited extent. Total reprogramming, for the moment at least, is limited to nuclei from very young embryos.

Why are older embryonic nuclei and adult nuclei less capable than younger embryonic nuclei of promoting normal development in enucleated egg cytoplasm? The answer may relate to a fundamental question in developmental biology: what mechanisms lead to the highly stable differentiated state of specialized cells in higher organisms? Modern cell biologists have accumulated evidence which suggests that brain cells are different from liver cells, not because they have different genetic material, but because their common genetic material is organized differently or utilized differently. The reduced yield of normal frogs from more differentiated (i.e., older and more specialized) nuclear donors reflects the highly stabilized condition of the differential organization or use of genetic material.

Many scientists have speculated about why adult nuclei seem less capable than younger nuclei of promoting normal development when transplanted to egg cytoplasm. One possible explanation is that DNA, the basic genetic material, may be altered or rearranged during the course of development (Brack et al., 1978; Fedoroff, 1984). The altered or rearranged genes *may* affect the capacity of a nucleus to be cloned. I do not know if changes in DNA structure are related to the failure to clone adult nuclei perfectly. However, if DNA changes in *many* differentiating cell types (and there is certainly evidence that it changes in *some*), then the cloning of adults (human and nonhuman) may be impossible for *genetic reasons*.

Another possibility is that the genetic material may be intact and whole in fully differentiated highly specialized cells, but that genetic material may be associated with other cell substances (substances that may convey stability to the differentiated state) so that the chromosomes of the transplanted nucleus are unable to respond to the rapid pace of cell division characteristic of the dividing egg. The inability of the nucleus and the cytoplasm to become synchronous in cell division (perhaps owing to differences in the replication timing of tissue-specific genes, see Goldmann et al., 1984) could (and does) lead to profound chromosomal disorders (DiBerardino and Hoffner, 1970; DiBerardino, 1979). Thus, nuclear-transplantation failure in adult cells may relate more to stability factors that ordinarily prevent inappropriate DNA replication than it does to structural changes in DNA. (Molecular mechanisms possibly involved in reactivation of previously dormant nuclei are considered in a review paper by DiBerardino, Hoffner, and Etkin, 1984.)

Microbiologists who study the controls of genetic activity in bacteria know far more about regulation than do scientists who work on similar problems in higher organisms. If more were known about gene regulation in higher organisms, it is possible that nuclei from maturing embryos and adults could be successfully cloned. Perhaps what keeps many nuclear transplanters at their laboratory benches is the belief that if they are ever able to devise a treatment for an adult nucleus so that it will be reprogrammed to direct development of a normal individual when cloned, then this mode of treatment will reveal much about the mechanisms of the differentiated state. And that is a primary concern of developmental biology.

CLONING AND CANCER BIOLOGY

Most readers know that many older people die of cancer. But many may not know that if a young person in the United States dies of a disease, the chances are greatest that the disease will be cancer. After several decades of intensive research, cancer remains a major medical problem in America.

Cancer related to environmental insult is receiving much attention. Monitoring air, water, and food for possible cancer-causing chemicals is becoming increasingly important with the recognition that noxious substances in our environment may cause cancer.

Although it is true that much cancer is related to environmental insult, it is also true that the origin of some (much?) cancer is not related to industrial chemicals, urban pollution, and cigarette smoking. Studies of fossil remains of pre-Columbian, Peruvian, and North American Indians show that a particular kind of bone cancer, multiple myeloma, was common (Morse et al., 1976). Certainly these unfortunate individuals were not exposed to automobile exhausts or chemical plant effluents. Such observations of cancer in prehistoric peoples suggest that even if the environment is cleaned of noxious human-made substances that are known or suspected to cause cancer, it seems probable that some cancer will still afflict humankind.

Cancer has been with humans for years and is not likely to go away in the near future. How then are those who become afflicted with the disease to be treated? There are three principal treatments: surgery, radiation, and chemotherapy. Chemotherapy is a particularly significant kind of cancer treatment because of the dismal fact that about two-thirds of cancer patients have metastatic cancer at the time of diagnosis. Metastatic cancer is cancer that has spread via the blood or lymph to many parts of the body, many secondary tumors forming as colonies from the primary cancer. Surgery cannot eliminate these myriad malignant growths, nor can radiation. Hence the need for chemotherapy (Pierce et al., 1978).

Unfortunately, chemicals produced to kill cancer cells are exceptionally toxic. They are toxic because it is difficult to design chemicals that will distinguish between a normal cell that is dividing and a malignant cell that is dividing. If more were known about cancer-cell genetic material and about regulating its differentiated state, there would be the rational hope that a cancer chemotherapeutic agent could be devised that would regulate differentiation of some kinds of malignant cells.

The urgent need for a nontoxic agent to regulate gene expression in malignant cells is one reason for doing cloning experiments designed to reveal the nature of the genetic material of the cancer cell and the capacity of that malignant genetic material to give rise to normal cell progeny. Cloning may be the most direct method of acquiring information about these two important aspects of cancer-cell biology.

The experimental design of one cancer-cloning study in which I participated is simple. Nuclei from a malignant tumor are inserted in a previously enucleated egg. "Malignant" refers to the capacity of a tumor to become invasive, to disseminate, and to colonize distant sites as metastases—because the Lucke kidney cancer is spontaneously invasive, disseminates, and becomes metastatic, it is malignant (Lucke, 1934b; Lucke and Schlumberger, 1949; McKinnell and Cunningham, 1982; McKinnell and Tarin, 1984).

If the recipient egg divides, one expects the result to be either a ball of cancer cells or a tadpole of some kind. A ball of cancer cells would suggest that the malignant nuclei being studied are not easily altered from their malignantly differentiated state. A tadpole would argue that the genetic material of the cancer under scrutiny can be redirected to a less malignant state.

If the tadpole is produced by cancer nuclear transfer, indicating a reversal from malignancy, further experiments could be suggested that might ultimately lead to control of malignant gene expression. It seems obvious to me that treating cancer by modifying gene expression promises to be less traumatic than treating cancer with cell-killing agents.

The leopard frog of North America, *Rana pipiens*, that served so well in the prototypic nuclear-transplantation experiments, may be afflicted with a kidney cancer (McKinnell, 1984) (Figure 4-2), which was formerly abundant (McKinnell, 1965) but recently has become less common (McKinnell et al., 1979; 1980). Dr. Balduin Lucke, late Professor of Pathology at the University of Pennsylvania, who originally described the frog cancer (Lucke, 1934a) believed that the kidney cancer cells had a normal number of chromosomes. It is now common knowledge that normal development demands normal

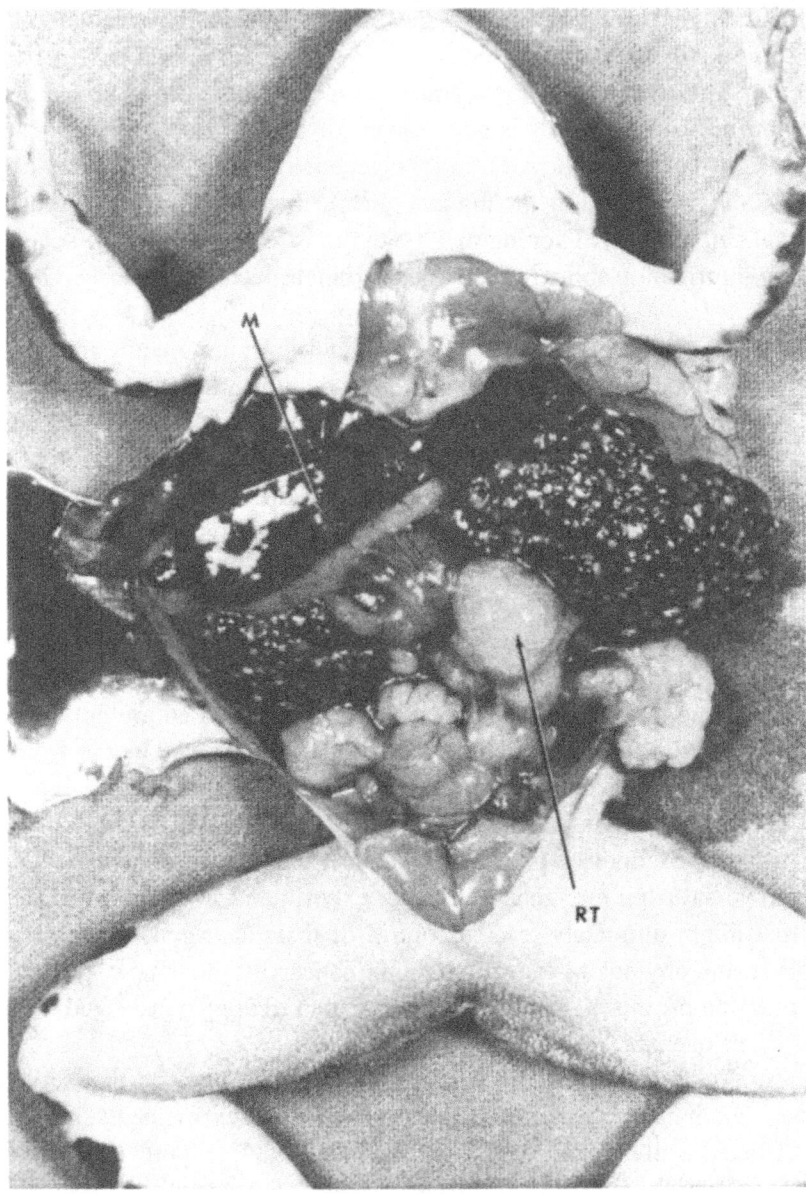

FIGURE 4-2. *An autopsy of a mature leopard frog,* Rana pipiens, *reveals large lumps on the kidney (RT). These masses are kidney cancer that has manifested its malignancy by spreading to the liver as a metastatic nodule (M). (From R. G. McKinnell, L. M. Steven, Jr., and D. D.* Labat, *"Frog renal tumors are composed of stroma, vascular elements, and epithelial cells: What type nucleus programs for tadpoles with the cloning procedure?" In N. Müller-Berát, ed.,* Progress in Differentiation Research, © 1976 North Holland, Amsterdam)

chromosomes—abnormal sets of chromosomes in humans result in abnormalities such as Down's syndrome (mongolism) (Lejeune et al., 1959). Therefore, the nuclear transplanter seeks cells with a normal chromosome constitution. The use of more sophisticated chromosome techniques enabled researchers to confirm Lucke's belief that the frog tumor was characterized by a normal number of chromosomes with normal form. Normal tadpoles and normal adult *Rana pipiens* have 26 chromosomes (Figure 4-3). The frog kidney cancer also has 26 chromosomes, and they appear similar or identical to those of the embryo and adult (Figure 4-4) (DiBerardino et al., 1963). The nuclear-transplantation procedure was perfected in *Rana pipiens*; nuclear donors should have normal chromosomes if normal or near-normal development is anticipated from the experiment; the frog cancer has a normal number of chromosomes—if any cancer was ideally suited for characterization by the cloning procedure, it was the frog kidney cancer.

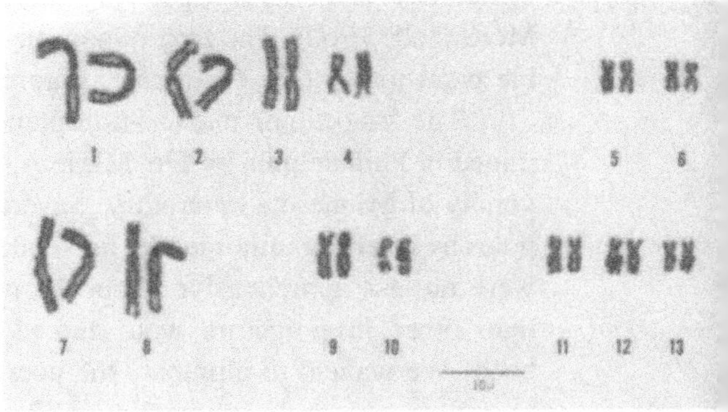

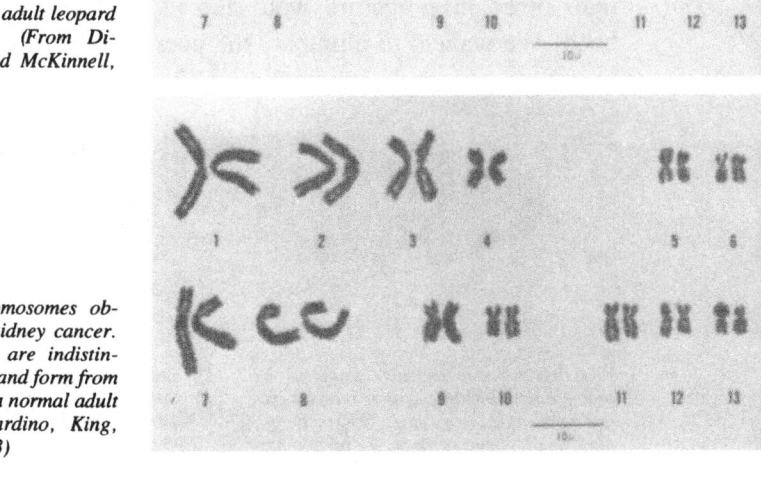

FIGURE 4-3. *Chromosomes obtained from a normal adult leopard frog,* Rana pipiens. *(From DiBerardino, King, and McKinnell, 1963)*

FIGURE 4-4. *Chromosomes obtained from a frog kidney cancer. These chromosomes are indistinguishable in number and form from the chromosomes of a normal adult frog. (From DiBerardino, King, and McKinnell, 1963)*

I collaborated with the co-developer of the nuclear-transplantation procedure, Thomas J. King, in studies of frog-tumor nuclear transplantation during the late 1950s. We found that instead of more cancer cells being formed when a cancer nucleus was placed in an enucleated egg, a tadpole was produced! That was an extraordinary observation in those early cloning days. Dogma said that cancer was an irreversible process. It was thought that a cancer cell could give rise only to more cancer cells. It had been assumed that the control for the highly stabilized malignant condition rested in the nucleus, which contained the genetic material. Thus nuclear progeny of a cancer nucleus would be malignant. However, instead of giving rise to more cancer nuclei, the malignant nucleus produced cell progeny that differentiated as nerve, muscle, gut, etc. These results demonstrated that at the very least those genetic components of an adult cancer nucleus that were required for forming an early embryo were present and still capable of functioning. Further, and perhaps more important, the results demonstrated that the differentiated state of cancer is not irreversible—at least frog kidney cancer is not (King and McKinnell, 1960). The frog tumor may be compared with a reversible plant tumor (see Chapter 1, Figure 1-2).

The frog-tumor nuclear-transplantation studies are being continued in Philadelphia by Dr. Marie A. DiBerardino and in my University of Minnesota laboratory. Several years ago, I produced tadpoles by inserting tumor nuclei into enucleated eggs (Figure 4-5). We were no less apprehensive about the possibility of parthenogenesis than other investigators who studied nuclear transplantation had been. We wanted to eliminate the possibility that the tadpoles were

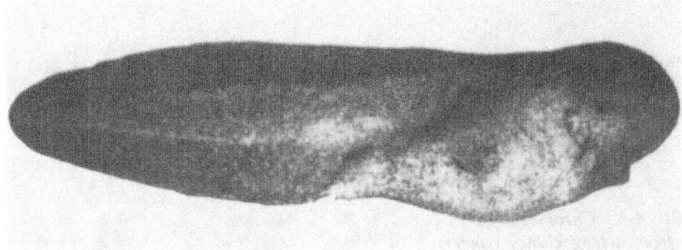

FIGURE 4-5. *A tadpole produced by transplanting a kidney-cancer nucleus into an enucleated frog egg. (From R. G. McKinnell, "Nuclear transfer in* Xenopus *and* Rana *compared." In R. Harris, P. Allin and D. Viza, eds., Cell Differentiation © 1972 Munksgaard International Publishers, Inc., Copenhagen, Denmark)*

produced with genetic instructions from maternal egg chromosomes and to establish that the development of each operated egg resulted from the programming of a cancer-cell nucleus in enucleated cytoplasm.

I referred earlier to the problem of parthenogenesis in the discussion of using donor nuclei tagged (i.e., labeled) with mutant-pigment-pattern genes (Kandiyohi and Burnsi). The female chosen to provide eggs did not carry the mutant gene, and expression of the Kandiyohi or Burnsi characteristic thus constituted genetic proof of the donor nuclei's participation in development. The pigment-pattern-mutant genes Kandiyohi and Burnsi were not suitable nuclear markers for the cloned tumor nuclei because the mutant genes are expressed only in the adult state—Kandiyohi and Burnsi tadpoles do not appear to be different from ordinary tadpoles—and the tumor nuclear-transplant embryos died as young tadpoles.

A tumor nuclear marker—frog cancer with an extra set of chromosomes—was designed and produced in my laboratory and in the laboratory of Dr. Kenyon Tweedell of the University of Notre Dame. Although it is true that an abnormal complement of chromosomes is incompatible with normal development in frogs and humans, it has been known for some time that frogs and salamanders can develop normally (except for a reduction in fertility) with an extra *set* of chromosomes. The normal leopard frog has 26 chromosomes per cell, 13 derived from the sperm and 13 from the egg. An embryo with 23 or 24 chromosomes (i.e., $26-3$ or $26-2$) develops abnormally. But a leopard frog with 39 chromosomes (26 from the egg and 13 from the sperm) develops normally! The 39-chromosome frog is normal in every way except that it has reduced fertility (Figure 4-6) (McKinnell, 1964).

A number of triploid embryos (embryos with three sets of chromosomes) were produced by hydrostatic pressure which keeps a diploid egg diploid even after fertilization. As noted earlier, we wanted triploid tumors because the 13 extra chromosomes would serve admirably as a nuclear marker or nuclear "tag." Most parthenogenetic embryos either have one set of maternal chromosomes (haploid) or remain diploid, as the egg is before fertilization. A triploid parthenogen would be an unlikely event. Thus triploid tumor nuclei should give rise to triploid embryos in a

FIGURE 4-6. *A triploid frog produced by the cloning procedure.
(From McKinnell, 1964)*

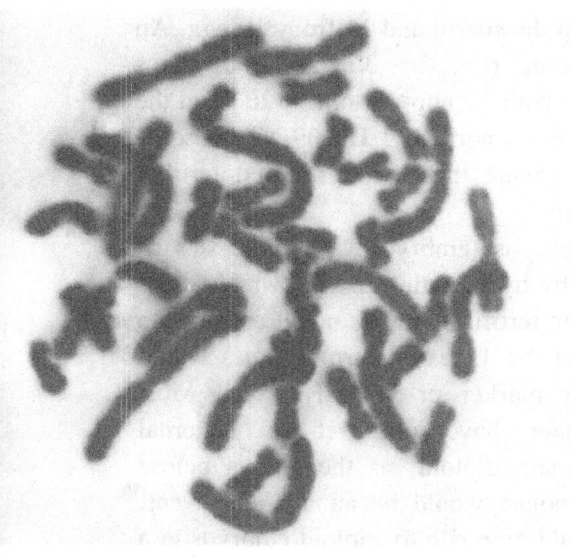

FIGURE 4-7. *Chromosomes of a
triploid kidney cancer of the northern leopard frog,* Rana pipiens.
*(Photograph courtesy of Dr. Marie
A. DiBerardino, from McKinnell
and Tweedell, 1970)*

true nuclear-transplant operation. Haploid or diploid embryos resulting from the insertion of triploid nuclei would unquestionably be the result of parthenogenesis.

To begin the experiment triploid tadpoles produced from pressure-treated eggs were flown to Notre Dame, where Tweedell provided a cancer-causing virus preparation to inject into the tadpoles. The tadpoles were returned to my laboratory where they developed into frogs. Some had cancer (McKinnell and Tweedell, 1970), and the cancers were different from any naturally occurring frog malignancies. They were triploid (Figure 4-7).

Triploid tumor nuclei were then transplanted into enucleated eggs. Seven tadpoles developed (Figure 4-8). The tadpoles swam. The seven swimming tadpoles had skin, connective tissue, muscle, brain, spinal cord, eyes, kidneys, liver, and all the other organs and tissues that characterize early tadpole anatomy. Swimming is significant because it demonstrates that not only are all of the requisite tissue types present and anatomically functional but the activity of the parts is coordinated (McKinnell et al., 1969a). What was the chromosome number of these seven creatures? All seven were triploid. There was no reasonable possibility that the tadpoles could have developed from any kind of nuclei other than the nuclei obtained from the triploid *cancer*.

Renal cancers of frogs contain connective tissue as well as cancer cells. Therefore, it could be asked if a connective tissue nucleus was transplanted instead of a cancer nucleus in these studies. It is highly unlikely that any nucleus other than a cancer-cell nucleus provides genetic material for the tadpoles. Why do I believe this? My colleagues and I examined cells from frog renal tumors under a microscope with an ultra-violet light source. The cells had been treated with a chemical, acridine orange, that caused them to fluoresce when viewed with ultra-violet light. This microscope procedure is similar to a procedure used to decorate some disco establishments. Black light (ultra-violet) shines on special paint and causes the paint to emit (fluoresce) bright colors of visible light. Cells can be treated so that they fluoresce, the glow revealing the *kind* of cell being studied. With fluorescence microscopy, we can distinguish between frog kidney-cancer cells and frog connective-tissue cells.

Several years ago, we wanted to ascertain what kind of cells are

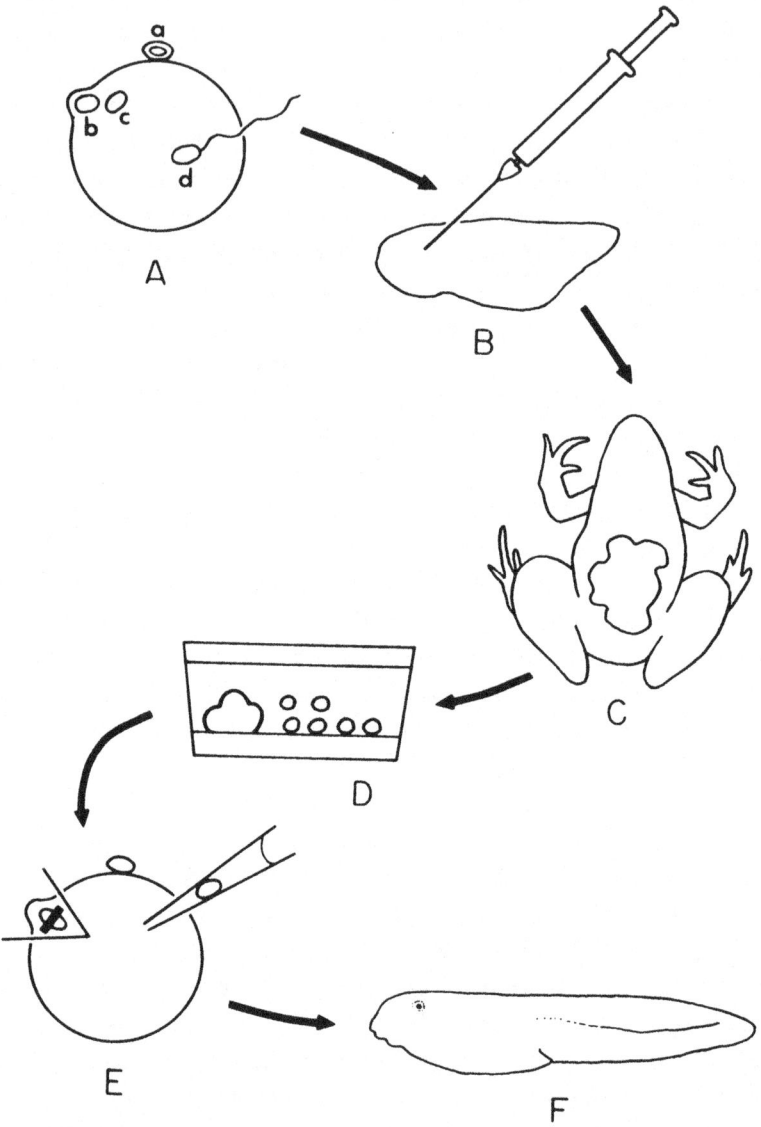

FIGURE 4-8. *Triploid tumor nuclear-transplantation diagram. Triploidy was induced by hydrostatic repression of the second polar body of inseminated eggs (A). Triploid embryos were then injected with a virus-containing tumor cell extract (B). The tadpoles developed into tumor-bearing triploid frogs (C). Triploid tumor cells were prepared for transplantation by dissociation in an electrolyte solution (D), transplanted into an enucleated egg (E), with a tadpole (F) resulting. The triploidy of the tadpole demonstrated the origin of the inserted nucleus, i.e., the tadpole developed as a result of the introduction of a tumor nucleus and not as a consequence of an inadvertently retained maternal nucleus. (From McKinnell and Ellis, 1972)*

present in a dish following the dissociation procedure we use to obtain donor nuclei for tumor-cloning experiments. Did the dish contain mostly connective tisue cells, mostly cancer cells, or a mixture of the two? This is an important question because the kind of cell present in the dissociation dish is the kind of cell that provides a nucleus for cloning. We noted that cancer cells were present in excess of 98% (McKinnell et al., 1976b). The study was repeated in my laboratory many times for a different reason but with the same result; almost all cells which detach from a tumor fragment are cancerous (Seppanen et al., 1984). We knew they were malignant because the cytoplasm fluoresced red. The cytoplasm of noncancerous connective tissue from frog renal tumors fluoresces *green*. Because the overwhelming majority of donor cells were malignant, we could be confident that the tadpoles described above were in fact derived from malignant nuclear donors.

The genetic (triploid) evidence and the cell study (ultra-violet microscopy) evidence show that the hereditary material of a frog cancer cell contains determinants for the formation of many cell types other than malignant ones. The tadpoles that developed from cancer nuclei also indicate that there is material in the egg cytoplasm that can coax or influence, in some as yet undefined way, malignant nuclei to give rise to nuclear progeny that include many seemingly *normal* types. If a frog had been produced in these cloning experiments, it would be evidence that the genetic material of a cancer cell is identical to the genetic material of a fertilized egg. Only tadpoles have been produced thus far with the cloning procedure using malignant nuclei.

The experiments raise questions: Do frogs not develop in cancer nuclear transfers because a virus is present? (Since the triploid tumors were formed by injection of a virus preparation, they presumably still contained virus.) Is the imperfect development attributable to the fact that the frog cancers were obtained from adults? There is no answer to these questions at present. Recall, however, that the malignant nucleus evokes as much development of the enucleated egg cytoplasm as do normal *adult* nuclei which have been similarly studied.

Nuclear transplanters are interested not only in the capability of specialized nuclei to be reprogrammed, with the formation of a tadpole, but, perhaps more, in what it is within the egg cytoplasm that calls forth the reprogramming. Thus there are studies being performed in Philadelphia on the capacity of special cytoplasm—the cytoplasm of *maturing* eggs—to enhance the developmental potential of differentiated nuclei. The molecular mechanisms responsible for enhanced gene reactivation in the nuclear-transplantation experiments, involving maturing (oocyte) egg cytoplasm, are not known, but they seem to involve stimulation of DNA replication in nuclei that are not dividing (DiBerardino and Hoffner, 1983).

Earlier, the need for new kinds of chemotherapy was stressed—the need for nontoxic cancer-curing substances that effect *differentiation* of cancer cells rather than death of cancer cells. I would like to suggest that characterization of the mechanisms by which egg cytoplasm, maturing or mature, is competent to activate dormant *normal* genes and to redirect differentiation from a previously malignant to a benign pathway may be a first step in the development of a nontoxic chemotherapeutic agent.

IMMUNOLOGY AND CLONING

Heart transplantation, developed in the 1960s by several surgeons trained at the University of Minnesota, is now so advanced that it is not a failure in surgery that causes the untimely death of many heart recipients. Rather, death results when the recipient rejects the newly inserted heart. Although surgical skills are adequate for tissue transplantation, the biology relating to rejection of foreign tissues is far less developed. It is apparent, therefore, that further knowledge of the immune response to tissue transplantation is urgently needed so that lives can be extended with existing surgical procedures.

Many animal species have been used in the study of how the immune response relates to tissue transplantation. Each species has its peculiar strengths and its distinctive weaknesses. Mice have been so useful in biomedical research that many strains with defined hereditary characteristics are available (Greenhouse, 1984). Since tissue rejection has a hereditary component, mice are useful because of what we know of their hereditary makeup. However, tissue transplantation is often studied with individuals that do not reject grafts of tissues or organs—and with mice, this means that grafts must be made between members of inbred strains. However, inbred animals lack vigor. Inbred mice are not a good model for humans, which are the products of near-random mating. There are social, religious, and legal taboos that minimize inbreeding in humans.

We know less about frogs genetically than we do about mice. However frogs, like humans, engage in random breeding within a population. They are perhaps the most studied species of vertebrate embryology. Cloning has been studied extensively in frogs. Because experimentalists have developed operative procedures, it is relatively easy to extirpate significant organs such as the thymus gland in a frog embryo—a significant gland because it is crucial in tissue rejection (Rollins-Smith and Cohen, 1982).

Can cloning be utilized in studies of tissue transplantation? To answer this question, the answers to two other questions are needed. First, do frogs respond to transplanted tissues with an immune reaction? Second, can frogs be obtained in groups that are as genetically similar as inbred mice but as vigorous as ordinary frogs?

The northern leopard frog, *Rana pipiens*, responds to foreign tissue (tissue obtained from unrelated individuals within the same species) grafts in much the same manner as humans. Frog skin grafted to an unrelated frog of the same species initially heals and flourishes. However, after a time, the graft dies and is lost. The death and loss of the skin graft is due to an immune response from the host. The rejection is quick and vigorous (Volpe, 1980). *Xenopus* also rejects grafted tissue, but the reaction is more sluggish and resembles that of lower vertebrates rather than humans.

Thus frogs have an immune response to grafted tissue. But are there groups of genetically similar or identical individuals? Certainly not naturally, for natural populations of frogs result from random mating. However, the cloning procedure can produce groups of genetically identical individuals.

Ordinary cell division gives rise to two precise replicates of the genetic material. When the fertilized egg divides, the two blastomeres that are formed are genetically identical. Human twins derived from a common egg are examples. The identical twins look alike and can even receive transplanted kidneys from each other without fear of rejection. Subsequent cell division, like the first division, results in genetically identical nuclear progeny. Thus all the cells of a blastula (and for the most part later stages, too) are thought to be composed of identical complements of DNA. Therefore if a number of frogs are produced by transferring nuclei from *one* blastula, the group should be as identical to each other as identical twins are.

I collaborated in a study with Peter Volpe, formerly of Tulane University, in the mid-1960s, in which I produced several groups of frogs by cloning. A common blastula served as nuclear donor for each cloned group, making each group isogenic. Isogenic means that each individual within the group had the same genes as the other members of the group. Volpe transplanted embryonic and juvenile frog tissue among members of the isogenic group, and there was no rejection reaction (Figure 4-9). Tissue grafted between individuals of *different* groups was invariably rejected. The simple experiment was useful in demonstrating that genetically identical groups of animals, which were not inbred, could be produced and that grafts were not rejected within these groups (Volpe and McKinnell, 1966).

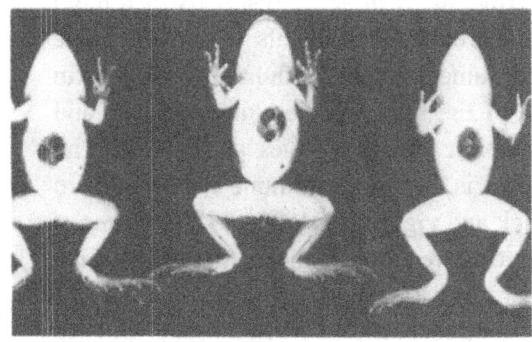

FIGURE 4-9. *Cloned frogs that are genetically identical do not reject patches of pigment-containing cells grafted to their bellies. (From Volpe and McKinnell, 1966)*

In most nuclear-transplantation experiments, there occur individuals with extra chromosome sets. They occur when the inserted nucleus divides before the egg cytoplasm divides. The two chromosome sets in one cytoplasm form two diploid nuclei. The diploid nuclei then fuse, forming one nucleus that is double the usual size and contains twice the ordinary amount of DNA. It is tetraploid. The tetraploid embryo becomes a tetraploid frog—since no new genes have been introduced in the isogenic groups, the tetraploid frog is qualitatively identical to the diploid frogs of the group. We already knew that the *quality* of genetic material affects the immune response; what we wanted to find out was whether the *quantity* of genetic material did so. The study of Volpe and McKinnell showed that tissue from a tetraploid individual was not rejected by isogenic diploid hosts and vice versa. This simple observation showed clearly that it is the diversity of genes, not the abundance, that is important in the rejection reaction. (For a review of immunobiology and cloning, see Volpe, 1980.)

Can cloning be useful in other immunobiological studies? I believe that it will be useful in studies of metastasis. Metastasis refers to colonization of a tumor at anatomical sites distant from the tumor's origin. The propensity to give rise to secondary tumor colonies can be measured experimentally by intravascular injection of tumor cells (Tarin and Price, 1979). The Lucke renal adenocarinoma, referred to previously, elicits an immune response when grafted to or injected into recipient frogs (Rollins and McKinnell, 1980; Tarin, McKinnell, and Nace, 1984). Consequently, if one wishes to study metastasis experimentally in ordinary frogs, immunosuppression with drugs is required to prevent the recipient frog from rejecting the administered tumor cells. Immunosuppression alters the host's biology in ways that may obscure the natural propensity for the tumor to grow metastatically in the host. Ideally, one would wish for the vigor of outbred animals with the capacity to accept tissue grafts. Isogenic frogs, produced by cloning, uniquely fill this order. We have begun a study, in collaboration with Dr. David Tarin of the John Radcliffe Hospital, Oxford University, England, using frogs as recipients of isogenic tumor. The protocol is exceptionally difficult and complex and will not be described here. What will be noted is the fact that isogenic frogs, produced by nuclear transplantation, have immunobiological characteristics that may prove to be of exceptional value in the study of the most malignant aspect of cancer—metastasis.

AGING—NEW INSIGHT THROUGH CLONING?

Aging may be the most important medical and human problem that can be served with the cloning procedure. As human populations become better able to contend with life-threatening medical problems such as infectious diseases, heart and kidney problems, and cancer, there is an increasing proportion of people who survive longer. At the same time, effective and widely used birth-control procedures result in fewer young people. It takes little imagination to contem-

71

plate the enormous social and economic problems of a reduced population of young people attempting to provide medical care for large numbers of incapacitated elderly.

Aging individuals who remain mentally alert and vigorously healthy support themselves at little or no cost to the rest of the population. In fact, they are a benefit to their community. They have time for the very young, they have vocational and professional experiences that can be a valuable resource to working people, and they are rich reservoirs of community and family history. However, aging people who are mentally and physically incapacitated are an emotional and economic burden. The enormous cost of caring for these unwell, aged people will escalate. If not for humane reasons (which are exceptionally compelling to me), then for economic reasons we need to know more about the aging process now.

Some scientists are concerned with extending the life-span—my personal view is that enhancing the quality of life of mature citizens is more urgent. Actually, the discussion of extension versus quality may not be very important, for they may well go hand in hand.

The scientific understanding of aging in the 1980s may be compared to the biologists' perception of cancer in the 1950s. That is, three decades ago it seemed that if biologists would only bend their skills to the problem of cancer, new insights would emerge that would likely result in a better control of the disease. Although a cancer cure has yet to be developed, much of the work that has been accomplished in the last three decades supports the confidence that the goal of cancer-cell biology will yet be achieved.

Today, extraordinarily little is known about aging. There are virtually no theories that marry cell biology, genetics, and biochemistry to an understanding of aging. Perhaps in the next few decades biologists can provide new insight for the new science of how organisms, and humans, grow old.

A number of aging experiments with cells cultured *in vitro* have been performed by Leonard Hayflick. The experiments suggest that normal cells have the potential for only a limited number of replications before they die. Normal cells are characterized by normal chromosome complements. Hence, according to Hayflick, cells with normal chromosomes have a limited life expectancy in culture (Hayflick, 1977). However, not all cells in culture have normal chromosomes. Years ago a woman died of cervical cancer. Before her death, some of her cancerous cells were placed in a culture medium. The cells continued to grow after her death and are still growing. The cells seem immortal. How do they differ from "normal" cells in culture with a limited life span? Careful chromosome analysis revealed that the cultured cells derived from the cervical cancer contained an abnormal allotment of chromosomes. It would seem, therefore, that normal cells in culture have a limited life expectancy and only malignant cells persist. An attractive aspect of the studies of Hayflick is that because normal cultured cells can endure only a limited number of cell divisions before they die, the changes that occur in the culture before the demise can be characterized. Senescence refers to changes associated with aging. Are the changes in cell cultures with normal chromosomes the result of the senescence of the culture? Are the changes brought about because of continuous cell division? Is cell division *per se* a principal cause of aging?

A skeptic might observe that what transpires in a culture flask may have little to do with what happens in an inact organism. Why? The environment of cells in culture is designed by humans. It is not uncommon to find a mixture of fetal calf serum or embryo extract, plus amino acids, and buffers as a growth medium. Antibiotics are frequently added to the medium. It is remarkable that cells survive in such a milieu. Perhaps the demise of the culture after a number of cell generations is the result of growing in a highly artificial medium. Perhaps metabolites accumulate or the cells have a limited capability of surviving in the synthetic environment. The apprehensiveness of the skeptic is only partly allayed by cell culturists who point out that cells derived from young individuals have the capacity to undergo more cell divisions than do cells derived from old individuals. Is there an alternative and feasible mode of examining the effect of many cell divisions on aging? If there is, can the data derived from the alternative mode be exploited to provide new and useful information about aging? The answer to that question is yes—and at least one alternative mode of study is, of course, cloning.

The time required for a complete cell cycle, i.e., the time from one cell division to the next, is very short during the early development of most amphibians and, of course, this includes *Rana pipiens*. When an embryo reaches the blastula stage, it has undergone 12 or

FIGURE 4-10. *Serial nuclear transplantation is a cloning procedure whereby nuclei may be induced to undergo many division cycles in a short period of time. The procedure may provide new information about aging.*

13 cell cycles and it is less than 1 day old. The duration of cell cycles increases as the embryo becomes older. As noted, a blastula nuclear donor has already undergone 12 or 13 cell cycles at the time it is inserted into enucleated egg cytoplasm. A nuclear-transplant blastula forms within a day after cloning and its cells have undergone 24 or 26 cell cycles. If the nuclear-transplant blastula serves as a nuclear donor, the donor nuclei will have sustained 24 or 26 cell cycles; when a blastula forms from this operation, the blastula cells will have experienced 36 to 39 cell cycles. The process of subcloning—also known as serial nuclear transplantation—can be continued day after day (Figure 4-10). Replication occurs with normal morphology of cells in a normal physiological environment. In theory, about 100 cell cycles could be induced by the end of day 8. It is known that serial cloning is feasible because it was first done, on a limited scale, by King and Briggs in 1956 and because we have already made extensive serial-cloning studies at Minnesota.

Hayflick reported senescence in cell cultures after about 50 population doublings. Although the actual number of cell cycles is greater than 50 in such cultures (cells lost during culture-medium change are not counted), it is clear that it will not take an inordinate effort to look for cell senescence among serially cloned embryos.

How would aging be characterized in serially cloned embryos? I do not know because the experiments are still in progress. Aging may not occur at all after serial nuclear transplantation. If aging fails to occur, it will be apparent that cell cycles *per se* are not responsible for senescence (at least in frogs). However, aging may be manifest in the serially cloned frogs in a reduced life span. Or it may be evident in an increased vulnerability to chromosomal aberrations, which may lead to a greater prevalence of abnormal development and tumor formation. There may be errors in DNA replication, or aging may be expressed as altered metabolism in the cloned frogs.

These comments are, of course, speculative. However, it is *not* conjecture to state that new information about the cell biology of aging can be generated with the cloning procedure. It is the business of the cloner, then, to use that information in conjunction with prospective studies to develop an understanding of aging that will benefit humankind.

5 Cloning Mice, Large Domestic Animals, and Humans

It has been reported that mice and some large domestic animals have been cloned. Humans have not. Because the reproductive biology of humans (and of other eutheria, i.e., mammals that have a well-developed placenta attached to the wall of the uterus) is similar to that of mice and other mammals, it is likely that humans could be cloned. The section on human cloning might be titled "Human cloning: The how to and the why not." This chapter presents the "how to"; it is basically a report on the capabilities of contemporary mammalian reproductive biology. The chapter concerns science, not opinion. The "why not" of human cloning is more a matter of opinion; it involves judgment and ethics. I have reserved my thoughts on the appropriateness of potential efforts directed toward human cloning for the Epilogue.

MICROMANIPULATION OF MUS MUSCULUS

The cloning of laboratory mice, *Mus musculus*, like the cloning of frogs, requires the following: the capability of obtaining eggs in sufficient number so that experiments can be performed; a means of egg enucleation for the production of host cytoplasm; a source of nuclei from donor embryos, and a means of combining the donor nucleus with the host cytoplasm. Mouse cloning, unlike frog cloning, requires an *in vitro* culture system which permits the synthetic embryo to develop to a stage for implantation into a surrogate (foster) mother. The experimenter then need only care for the surrogate mother until birth of the clone (Illmensee, 1984).

It is fashionable to believe that scientific achievement and the rate that scientific knowledge advances is a function of scientific equipment. There is, of course, some truth to this notion. Imagine, if you will, trying to understand the anatomy of a virus without an electron microscope. However, I have a strong impression that the spirit of the scientist in quest of new understanding is far more important to the advancement of knowledge than is scientific hardware. Frogs were cloned first by Briggs and King in Philadelphia in 1952 (Chapter 3). There were no new mechanical devices that made these experiments possible—rather, it was the questing spirit of two pioneer investigators. Similarly, all the procedures requisite for mouse cloning predate the 1981 and 1983 reports of that feat. For the cloning of mice, no new equipment was sought or needed. What was essential was the determination to persist until success rewarded the endeavor. The discussion that follows relies heavily on the publications of Illmensee and Hoppe (1981); Hoppe and Illmensee (1982); and McGrath and Solter (1983a; 1983b; 1984a), who have reported successful mammalian nuclear transfer. Reports of earlier attempts were made by Bromhall (1975).

Eggs for Experiments

The ova used for host cytoplasm in mouse cloning were different from the ova used in frog cloning. Enucleated *unfertilized* eggs were used in frog experiments. Successful mouse cloners have, thus far, used as recipient cytoplasm enucleated *fertilized* ova before the fusion of the male and female pronuclei ("pronuclei" are vesicles which contain the genetic material of the sperm and of the female that furnished the egg). The pronuclei were removed surgically.

Male and female mice were permitted to mate naturally. In mice ovulated eggs are released from the ovary and move toward the uterus via the oviducts (also known as the Fallopian tubes). Sperm, deposited at the time of mating, travel through the uterus and up into the oviduct, where, if eggs are available, fertilization takes place. Shortly after fertilization, the oviduct is surgically removed and an incision is made in it. Fertilized eggs, which were under pressure in the oviduct, readily flow out at the site of the incision (Biggers et al., 1971).

The mouse egg, like the eggs of other mammals, develops in a blister-like follicle of the ovary. Some of the follicle cells, which persist and remain attached to the egg after ovulation, form a cloud of tiny cells that, to some extent, hides the relatively large egg (Figure 5-1). The cloud of follicle cells heaped up around the egg is known, appropriately, as the cumulus. Because the structure of the

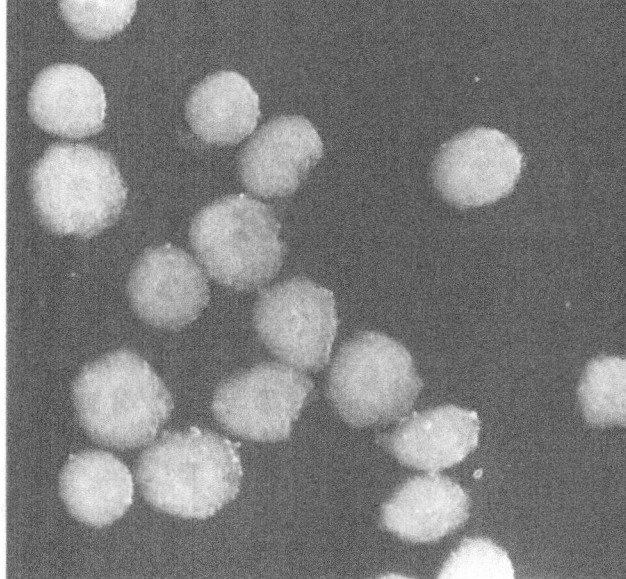

FIGURE 5-1. *Freshly ovulated ova of the short-tailed shrew,* Blarina brevicauda. *The eggs are partially obscured by a cloud of cumulus cells. The shrews were wild-caught in Minnesota, ovulated in the laboratory, and their ova were photographed by S. Goustin, R. Stanek, and R. Werner, former undergraduate students of the author.*

fertilized egg is obscured by the cumulus, the cumulus cells of ovulated eggs must be dispersed before any sort of micromanipulation can be performed.

Several enzymes are available that digest intercellular substances resulting in cell dispersal (Figure 5-2). One such enzyme is

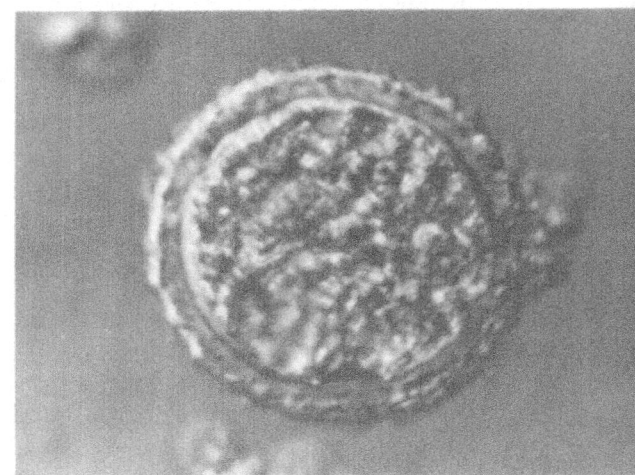

FIGURE 5-2. *An ovum of the short-tailed shrew,* Blarina brevicauda, *with cumulus cells removed by enzyme treatment. Photograph by S. Goustin, R. Stanek, and R. Werner, former undergraduate students of the author.*

hyaluronidase. Treating the fertilized egg and its cumulus cells with this enzyme results in the liberation of a clean and naked zygote (Brinster, 1971).

The number of eggs ovulated can be increased with hormone treatment (Fowler and Edwards, 1957; Biggers et al., 1971). Superovulation occurs when young adult mice are injected with the hormones known as pregnant mare serum and human chorionic gonadotropin. Twenty or 30 ova per mouse is the yield with hormone injection. Even more ova can be obtained with superovulation of immature mice (Gates, 1971).

Enucleation of Host Cytoplasm

The fertilized egg's pronuclei may be removed through microsurgery. It is reported that eggs survive microsurgery when cultured briefly in the drug Cytochalasin B, or that drug plus another drug, colcemid (Illmensee and Hoppe, 1981; Markert and Seidel, 1981; McGrath and Solter, 1983a). Cytochalasin B relaxes cytoplasmic microfilaments (Luchtel et al., 1976), while colcemid causes dissolution of cytoplasmic microtubules (Mareel and DeBrabander, 1978). The effect of one or both of the drugs is to reduce the egg's cytoplasmic rigidity, which facilitates cell surgery.

Enucleation, or any cell surgery, is not feasible unless the egg is held secure in some device. A suitable egg-holding device is a blunt and polished pipette (Figure 5-3). Gentle suction is applied via the holding pipette to secure and position the ovum (Hoppe and Illmensee, 1977). The holding pipette and its egg are positioned under a microscope with the aid of a micromanipulator.

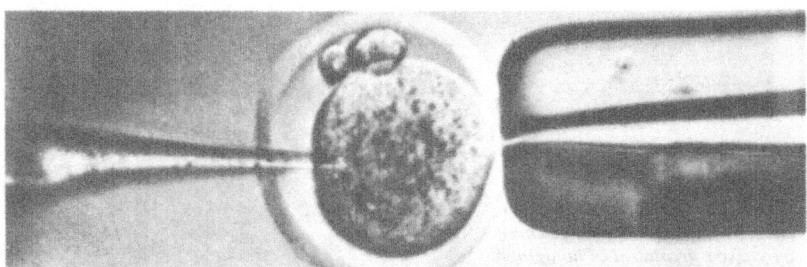

FIGURE 5-3. *A mouse egg is held for microsurgery with gentle suction applied* *via a blunt and polished holding pipette (shown on right).*

FIGURE 5-4. *Enucleation and nuclear transplantation into a fertilized mouse egg. (A) The fertilized egg with both pronuclei visible (arrows) is held in position with a holding pipette and the zona pellucida is slightly depressed by the sharp nuclear-transfer micropipette (left) containing an inner-cell-mass nucleus. (B) The transplanted nucleus is inserted and the micropipette is moved near the male pronucleus (small arrow) for removal. (C) Later, the female pronucleus (large arrow) is removed by sucking it into the same micropipette. (D) After removal of both pronuclei, the micropipette is gently removed. (From K. Illmensee and P. Hoppe, "Nuclear transplantation in* Mus musculus: *Developmental potential of nuclei from preimplantation embryos,"* Cell, *1981, 23:9-18. Copyright © 1981 by MIT)*

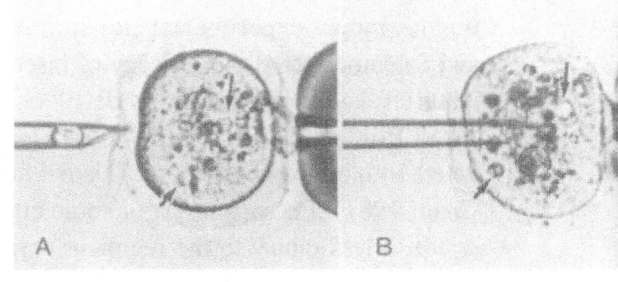

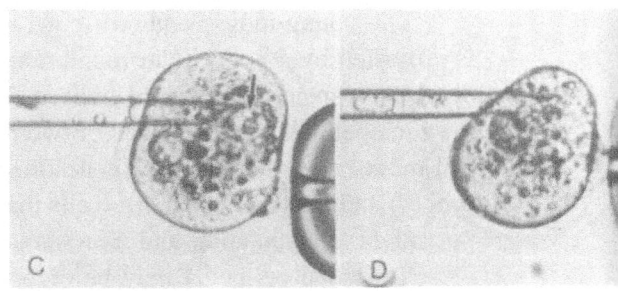

Two modes of enucleation have been reported in successful mouse nuclear transplantation. In one study, a sharpened micropipette containing the donor nucleus was inserted into the previously prepared recipient fertilized egg. The donor nucleus was released. Leaving the micropipette in the ovum, the experimenter sucked the male and female pronuclei into it, after which the pipette was removed from the ovum (Figure 5-4). One surgical invasion with the micropipette accomplished both nuclear transfer and enucleation (Illmensee and Hoppe, 1981). An even gentler surgical procedure was reported by McGrath and Solter (1983a). These investigators enucleated fertilized eggs *without* penetrating the cytoplasm. A micropipette was placed on the surface of the egg adjacent to a pronucleus. Gentle suction resulted in the aspiration of cytoplasm containing a pronucleus. The cytoplasm was connected to the egg with a slender cytoplasmic bridge. When the bridge was severed, the ovum was deprived of the pronucleus without having been penetrated by the micropipette. The other pronucleus was similarly removed.

Removal of pronuclei by surgery is, of course, not the only option for mouse experimental cloning. Several years ago, my students and I demonstrated the efficacy of laser ablation of frog-egg genetic material as a noninvasive mode of enucleation (McKinnell et al., 1969; Ellinger et al., 1975). Mouse embryo nuclei have been inactivated with a microbeam laser (Daniel and Takahashi, 1965; Lin and Chan, 1981). Perhaps laser, or some other form of irradiation, could result in less injury to the recipient egg with the beneficial result of a greater yield of cloned mice.

Preparation of Nuclei from Donor Embryos

The young mouse embryo (e.g., an embryo about 4 days old) is protected by a noncellular membrane, the zona pellucida (Figure 5-4) which encloses the entire embryo. The embryo contained within the zona pellucida is composed of two primary cell types: the inner-cell mass (a population of cells destined to give rise to the embryo proper) and the trophectoderm (cells that will give rise to tissues important for implantation and nourishment of the embryo).

The zona pellucida may be digested with an enzyme (Mintz, 1971) to expose the embryo for cellular dissociation. Or the zona can be removed mechanically (Illmensee and Hoppe, 1981). The zona-free naked embryo can then be separated into the inner-cell mass and trophectoderm with sharpened, very fine, metal needles. Dissociating the inner-cell mass into free unattached cells can be accomplished with enzymes. The enzymes digest the material that ordinarily permits cells to stick together. A calcium- and magnesium-free buffered salt solution containing protein-digesting enzymes produces complete dissociation. It can be hastened by repeated swirling into and out of a pipette.

Nuclear Transplantation

Conceptually the simplest way to place a nucleus into egg cytoplasm is to insert it with a micropipette. The procedure works. Illmensee and Hoppe (1981) sucked a previously dissociated donor cell into a micropipette smaller than the cell. The donor cell was broken by the small diameter micropipette in such a manner that the nucleus was liberated and remained undamaged. The donor nucleus was then placed in recipient egg cytoplasm before the egg was enucleated.

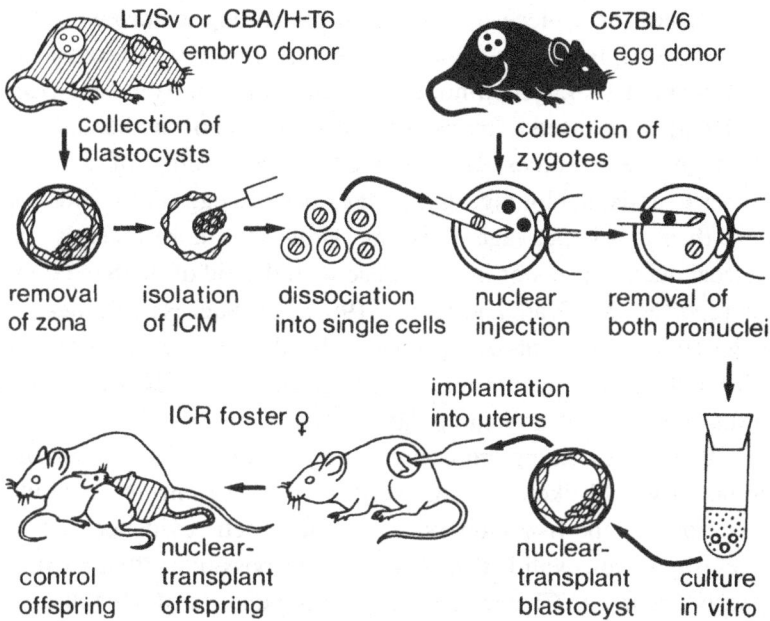

LT/Sv or CBA/H-T6 embryo donor

C57BL/6 egg donor

collection of blastocysts

collection of zygotes

removal of zona · isolation of ICM · dissociation into single cells · nuclear injection · removal of both pronuclei

implantation into uterus

ICR foster ♀

control offspring · nuclear-transplant offspring

nuclear-transplant blastocyst · culture in vitro

FIGURE 5-5. *Nuclear transplantation in the mouse. Donor embryos were collected from females on day 4 and kept in a culture medium. After surgical removal of the zona pellucida, the blastocysts were dissected manually into the inner cell mass (ICM) and trophectoderm (TE), which were then dissociated enzymatically into single cells. An ICM or TE cell was disrupted by sucking into a small glass pipette, and the cell nucleus with surrounding cytoplasm was injected into a fertilized egg. Following nuclear injection, the* genome *of the recipient egg was removed by sucking the male and female pronucleus into the micropipette. The nuclear-transplant embryos were cultured to the blastocyst stage and then transferred together with some control embryos into the uterus of a pseudopregnant foster female in order to allow development to term. (From K. Illmensee and P. Hoppe, "Nuclear transplantation in* Mus musculus: *Developmental potential of nuclei from preimplantation embryos,"* Cell, *1981, 23:9-18. Copyright © 1981 by MIT)*

Enucleation followed. (Figure 5-5.) Although the sequence differs from that of the frog nuclear-transplantation technique, the end result is the same.

Can nuclear transplantation be accomplished *without* the trauma of surgical nuclear implantation? Spontaneous fusion sometimes occurs when cells of two types are cultured together (Barski et al., 1961; Weiss, 1980). The Sendai virus, added to a culture medium, increases the rate of cell fusion many hundreds of times (Okada, 1958; Solter, 1981). The virus virulence is reduced by ultraviolet irradiation or chemical inactivation (Watkins, 1973). Can a virus effect fusion of a piece of cytoplasm containing a nucleus with an egg? Mouse spleen cells (Graham, 1971), fibroblasts (Baranska and Koprowski, 1970), lymph-node cells, and bone-marrow cells (Lin et al., 1973) were fused to mouse eggs with the aid of the Sendai virus. Although these studies demonstrated the possibility of exploiting viruses to assist in cloning, in fact, the operated egg rarely underwent more than a few cleavage divisions. However, more recently, non-surgical nuclear transfer in the mouse with the aid of the Sendai virus was reported (McGrath and Solter, 1983a, b). Success was measured in the survival of embryos produced by the procedure (more than 90%) and development to term at a frequency ''not significantly different from that of nonmanipulated control embryos.''

Before describing embryo culture and transfer to the surrogate mother, I would like to remind readers of the extensive, meticulous work done by the frog cloners to provide genetic evidence that their conclusions were valid, that there was no possibility that parthenogenesis occurred (Chapter 3). Micromanipulators of *Mus musculus* have been no less diligent. Appropriate genetic measures were taken to ensure the validity of the experiments.

In Vitro Culture of Mus Musculus

Producing an egg with an implanted nucleus, after which the egg is deprived of its own genetic material, while an important accomplishment, is only the first step in the production of a cloned mouse. The egg must be cultured in an artificial medium until it has developed to an embryonic stage suitable for implantation into a prepared foster mother. This *in vitro* development in the laboratory replaces the time that ordinarily elapses between fertilization in the oviduct and arrival of the embryo in the uterus where implantation and fetal development occur.

Culture of mammalian embryos *in vitro* may have the sound of science at the cutting edge of technology, but, in fact, it has a relatively long history. The first successful experiments involving mammalian embryos cultured *in vitro* were recorded about three-quarters of a century ago by Brachet (1912). In the past several decades, results with *in vitro* culture (Biggers et al., 1971; Sanyal and Naftolin, 1983; Ackerman et al., 1984) have improved to the extent that survival and growth of early mouse embryos for a period of 10 days (about half of the gestation period) is now possible (Chen and Hsu, 1982). The *in vitro* cultured mouse embryos have limb buds, liver, pancreas, lungs, brain, beating hearts, etc.

The nuclear-transplanted mouse embryos in the successful mouse-cloning experiments were cultured *in vitro* 4 or 5 days (Illmensee and Hoppe, 1981; McGrath and Solter, 1983).

Implantation of Cultured Cloned Embryos into Uteri of Foster Mothers

Although the technology of *in vitro* culture has made remarkable strides, it is not yet possible to culture a mammalian embryo for the entire fetal period. Thus, to obtain a cloned mouse, the embryo must be implanted in the uterus of a receptive female, which becomes the foster mother of the clone. At birth, the foster mother has the added chore of a wet nurse. The transplantation of young embryos to foster mothers is not a new procedure either. The uterus of one variety of rabbit was used for the nourishment, growth, and complete fetal development of another variety of rabbit just a little less than a century ago (Heape, 1890). Since that time, foster-mothering of this type has been extended to the rat, hamster, sheep, pig, goat, and other species (Dickmann, 1971; Kraemer, 1983) including, recently, the water buffalo (Drost et al., 1983) and *human* (Franklin, 1984). Embryo transfer is so successful in cattle that it has become a rapidly growing agricultural business (Seidel, 1981) (Figure 5-6). Mouse

FIGURE 5-6. *The cow (upper right) is the genetic parent of the 10 calves in the foreground. The cow was superovulated and embryos were obtained from her 7 days after fertilization. The embryos were cultured briefly* in vitro *and then transferred to the uteri of the dark cows in the background, where they developed to term. (Photograph courtesy of Dr. George E. Seidel, Jr.)*

embryo transfer to the uterus of a foster mother was described some years ago (McLaren and Michie, 1956).

Transplantation of an early embryo to a uterus is simple in principle: the embryo is transferred surgically by making an incision through the body wall with exposure of the uterus, or, by a nonsurgical means. For nonsurgical transplantation, the transfer pipette containing the embryo (nuclear-transfer embryo in the present situation) is inserted through the birth canal past the cervix, and the embryo(s) are deposited in the cavity of the uterus (Moler et al., 1979). The foster mother is pseudopregnant. This means that she was mated with a sterile (vasectomized) male, and she has the physical appearance of early pregnancy. After embryo transfer, she is no longer pseudopregnant—she is truly pregnant. Some 20 or 21 days later cloned mice are born. In some cases they are reared to adulthood (Figure 5-7) and sexual maturity.

Yield of Mice with the Cloning Procedure

The first published study of cloned mice reported that 142 of 363 (39%) eggs survived enucleation and the transfer of inner-cell-mass nuclei. About 25% of the survivors cleaved (i.e., the experimental egg underwent cell division), and 13% were cultured *in vitro* and reached a developmental stage just before that required for im-

FIGURE 5-7. *Mice produced by transplanting inner-cell-mass nuclei from blastocysts into enucleated eggs. (From K. Illmensee and P. Hoppe, "Nuclear transplantation in* Mus musculus: *Developmental potential of nuclei from preimplantation embryos,"* Cell, *1981, 23:9-18. Copyright © 1981 by MIT)*

plantation in the uterus. Sixteen (4%) of inner-cell-mass nuclear-transplant embryos were implanted into the uteri of pseudopregnant females. Three (0.8%) live-born mouse clones resulted (Illmensee and Hoppe, 1981). The report was extraordinarily important because it indicated that mice could be produced by the nuclear-transplantation procedure. Refinements in the technique, to increase yield, could be attended to now that successful cloning had been demonstrated.

Four cloned mice (3.7%) were born from a study of 107 eggs transplanted with parthenogenetic inner-cell-mass nuclei (Hoppe and Illmensee, 1982). The significance of the parthenogenesis study is discussed below.

It is difficult to compare the study of nuclear transplantation by surgical insertion with nuclear transplantation by fusion with the Sendai virus primarily because the nuclei transplanted by virus fusion were pronuclei obtained from one-cell fertilized eggs. Fertilized egg nuclei are, of course, totipotent. Thus, the procedure as reported does not provide any information concerning nuclear differentiation. It does, however, provide an alternative technical method for cloning. Sixty-four nuclear-transplant embryos were transferred to the uteri of pseudopregnant females. Ten mice (16%) were born; seven survived to adulthood.

Totipotent Mouse Nuclei

Only nuclei from frog embryos have been shown to be totipotent (DiBerardino, 1980; McKinnell, 1981). No adult frog has ever been cloned from an adult nuclear donor (Chapter 4). It is entirely too early to judge whether or not a similar situation exists in mice or other mammals. As was stated in the Preface, there is some disagreement concerning results of mouse nuclear transplantation. However, there are suggestions that mouse nuclear transplantation may be similar to frog nuclear transplantation. The similarity relates to early loss of nuclear totipotency as revealed by nuclear transfer. The capacity of amphibian embryonic nuclei to promote normal development, when transplanted to an enucleated ovum, is rapidly lost after the onset of gastrulation (see Figure 4-1 and related discussion, Chapter 4). Although there was controversy initially among amphibian embryologists regarding the chronology of totipotency loss when comparing *Xenopus* with *Rana* nuclear transfers, there is now no reason to doubt the validity of the data shown in Figure 4-1 (see also McKinnell, 1978).

At the present time, however, there is considerable controversy about the loss of totipotency in mouse nuclear transfers. Illmensee and Hoppe (1981) reported that at least some inner cell mass nuclei of blastocysts are totipotent; trophectoderm nuclei are not. Their study seems to be supported by Modlinski (1981), who used a different procedure but obtained essentially the same result. Modlinski surgically transplanted inner cell mass and trophectoderm nuclei to *non*enucleated eggs—Illmensee and Hoppe enucleated the eggs—and the zygote nucleus fused with the transplanted nucleus. Trophectoderm nuclear transplants died after only a few cell divisions, but inner cell mass nuclear transplants developed normally *in vitro* to the blastocyst stage. Modlinski's results suggest early loss of developmental potential in trophectoderm but retention of that potential in inner cell mass nuclei—results in harmony with those reported by Illmensee and Hoppe.

More recently, McGrath and Solter (1984b) attempted to transplant nuclei from mouse embryos obtained from cleavage stages through blastocyst stages. They reported continued success with the transfer of nuclei from the zygote (as reported in McGrath and Solter, 1983a, 1983b, 1984a), but, unexpectedly, they reported a precipitous decline in developmental capacity as early as the two-cell stage. No four-cell, eight-cell, or inner cell mass nuclei were competent, in their laboratory and with their procedure, to program for normal development. Nuclei from the eight-cell stage or inner cell mass blastocyst stage, transplanted to fertilized eggs which were subsequently enucleated, resulted in embryos that stopped developing after only one or two cell divisions. McGrath and Solter (1984b) attempted 231 transfers of inner cell mass nuclei by *surgical* implantation. No recipient egg developed as far as the blastocyst in these experiments.

Thus, there is disagreement concerning results obtained with nuclear transfer in mice. Is this disagreement the result of differences in procedure, in skill, or is it due to some other factor or group of factors? I do not know. Mammalian cloning is in its infancy. The controversy about mammalian nuclear-transplantation results will undoubtedly be resolved as the discipline matures.

Parthenogenesis in Mice—a Cloning Study

Parthenogenesis is development of the unfertilized egg. It is well known and has been extensively studied in a diversity of animals, including sea urchins (Sachs and Anderson, 1970) and frogs (Parmenter, 1933). There are natural populations of parthenogenetic lizards (Moritz, 1983; Cole, 1984). Despite the occurrence of parthenogenesis in disparate species, it is not usually associated with mammals. However, a strain of mice known as LT gives rise to eggs which start developing parthenogenetically. The eggs undergo cell division after ovulation and implant in the uterus without the benefit of interaction with sperm. Shortly after implantation, however, the parthenogenetic embryos die and are aborted.

Does abortion occur because the eggs are genetically defective or because of an essential change that occurs in an egg at fertilization? Parthenogenetic cells (obtained from LT embryos before uterine implantation) were combined with genetically different normal embryonic cells in an effort to ''rescue'' the parthenogenetic cells. The combination results in a *viable* embryo. The ''rescue'' experiment established that parthenogenetic mouse cells are capable of participating in normal development (Stevens, 1978; 1980).

The study suggested that lethality of young parthenogenetic mice might be due to something other than genetic defect. However, this was not entirely clear because the cells from the normal embryos may have provided essential genes missing from the parthenogens. A direct test of whether or not genetic defect exists in parthenogenetic mice was possible by nuclear transplantation. Nuclei from parthenogenetic embryo inner-cell mass were transplanted into fertilized ova of a different genetic strain. The recipient ova were enucleated after nuclear transfer as described above. Four mice were born. All were of the parthenogenetic donor genetic type; none expressed genes of the recipient ovum (Hoppe and Illmensee, 1982). Lethality of LT mouse parthenogens was not attributed to genetic defect. The parthenogenetic nuclei were reported to be totipotent, which means that they had all of the genes necessary for complete and normal development.

Surani et al. (1984) reported that parthenogenetic haploid eggs develop normally when restored to the diploid condition with a *male* pronucleus (transplanted after the method of McGrath and Solter). Development continues to sexual maturity. Reconstitution to the diploid state of a parthenogenetic haploid egg with a *female* pronucleus results in abnormalities. Because the *entire* genome of the Hoppe and Illmensee experiment is derived from the female, Surani et al. consider their results to differ from those of Hoppe and Illmensee.

The procedures and mouse strains are, of course, quite different. One group transplanted nuclei of spontaneously occurring diploid parthenogenetic embryos, the other utilized haploid parthenogens which were restored to diploidy with a transplanted male or female haploid pronucleus. One group used surgical transplantation, the other used a virus to fuse pronuclei. Resolution of these seemingly inconsistent results awaits further experimentation.

Mouse Cloning: Where to from Here?

The scientific literature on mouse cloning is scant. What does the future hold for the procedure? It seems to me that before extensive studies are designed to exploit mouse cloning, many additional studies are necessary to establish optimal experimental procedures. One laboratory reported producing mice by surgical transplantation of a donor nucleus. Another laboratory fused the donor nucleus to the recipient egg with the aid of a virus. Neither laboratory has published extensively nor has had long experience producing mice by nuclear transfer. Thus, I believe there will be a period of time devoted to experiments designed to obtain optimal results. After that, perhaps the most interesting problem to study will be differentiation. Will there be a progressive decline in the capacity of nuclei from body cells to program and direct normal development—as has been reported in frog nuclear-transplantation experiments? Can that restriction (if it occurs) in developmental potential be modulated by pretreatment of the donor nucleus before nuclear transplantation?

A pretreatment that might enhance the competence of a mouse embryonic nucleus to program for normal development is conditioning by oocyte cytoplasm. DiBerardino and Hoffner (1983) reported that nuclei from adult frog red blood cells, after a 24-hour exposure to oocyte cytoplasm, have the capacity, when transplanted, to direct the formation of tadpoles. Red blood cell nuclei not pretreated by exposure to oocyte cytoplasm develop no further than abnormal blastulae, or with retransplantation to gastrulae. Oocyte cytoplasm seems to reactivate dormant genes in frog red blood cell nuclei. Although the *mechanisms* for gene reactivation by oocyte cytoplasm are not known, the *effects* are substantial (DiBerardino and Hoffner, 1983; DiBerardino et al., 1984).

Mouse embryos are reported to precociously express their own genes, in some instances as early as the two-cell stage (Sawicki et al., 1981; Magnuson and Epstein, 1981; Johnson, 1981). This early specialization may reduce the capacity of mouse nuclei to participate in the normal development of a nuclear transplant. It, therefore, seems reasonable to suggest that mouse embryonic nuclei, like frog nuclei, may respond to the conditioning influences of oocyte cytoplasm.

Clearly, the mouse is a better laboratory animal than the frog from the point of view of genetics. There are hundreds of mouse inbred strains and mutant genes (Staats, 1980; Greenhouse, 1984). Many of the mutations affect biochemical pathways that have been studied extensively. Mouse cloning to provide new insights into developmental, genetic, and biochemical biology awaits creative exploitation.

EXPERIMENTAL TWINNING IN LARGE DOMESTIC ANIMALS—A FORM OF CLONING

Blastomere separation of sea urchin embryos and amphibian embryos was described earlier (Chapter 2). More than one complete sea urchin or amphibian developed from the cell-separation experiments. The production of more than one embryo from separated cells of a single embryo, without the intervention of sex, is by definition a form of cloning. From a historical point of view, blastomere separation is an older cloning method than nuclear transplantation. Nevertheless, the procedure effectively produces mice and large domestic animals.

Mice are small and relatively easy to work with. It is not surprising, therefore, to note that attempts have been made to produce two identical young from a single mouse embryo. Embryos comprised of two or more cells have been bisected, cultured *in vitro* for

24 or more hours, and transferred to recipient foster mother mice. Live embryos and young have ensued from the half embryo experiments (Moustafa and Hahn, 1978; Tsunoda and McLaren, 1983), results that are in complete harmony with the classic sea urchin and frog blastomere separation studies.

It would be useful to know if the procedures used to produce mice twins could be extended to large mammals, many of which have agricultural importance. Bisection of embryos with subsequent survival and growth has been successful in several domestic species.

Two-cell sheep embryos were removed from their protective zona pellucidae, the blastomeres were separated mechanically, and the individual separated blastomeres were placed into previously emptied zona pellucidae. Since the recipient zonae were opened to permit the extraction of the resident embryo to make room for the inserted experimental blastomere, the zonae had to be sealed in some manner to prevent loss of the inserted blastomere. The zonae with their blastomeres were placed in liquid agar (a substance that will form a soft gel). The agar served as a protective envelope around the operated zona pellucida and its blastomere. The gel-coated zona and its contents were then placed in a larger agar gel to form a more manageable particle. The double-coated zona and its blastomere were then implanted into a foster mother. Several pairs of monozygotic lamb twins have been born (Willadsen, 1979) (Figure 5-8). Similarly, twins ensue from the separation of 4- and 8-cell sheep embryos (Willadsen, 1981).

Early experimenters placed the split embryo and its zona pellucida into an agar coating, as described above. The agar treatment is no longer considered necessary by some reproductive biologists. Good results have been obtained by placing the half embryo into a zona pellucida and transferring it immediately into a recipient female (Lambeth et al., 1983).

The same procedure has been used to produce monozygotic twins, triplets, and even quadruplets from bovine embryos. The triplets and quadruplets, of course, were derived from embryos that had developed beyond the two-cell stage. Three genetically identical female calves, produced from micromanipulation of an 8-cell embryo, revealed expected similarity in hair pigmentation patterns. A fourth embryo of this isogenic group of three died after day 50 of

92

pregnancy (Willadsen and Polge, 1981).

FIGURE 5-8. *Sheep blastomere separation at the two-cell stage resulted in the three sets of twins shown here. Pigmentation patterns are very similar but not identical in these laboratory-produced mono-zygotic twins. (From Willadsen, 1981. Reproduced by permission from* Nature *277:298-300, copyright © 1979 Macmillan Journals Limited.)*

Perhaps even more remarkable than the separation of blastomeres in young embryos composed of only a few cells is the surgical bisection of 8-day-old bovine embryos. Embryos of this age have already differentiated the specialized trophoblast cells destined to nourish the developing fetus and the inner-cell mass, destined to become the embryo proper. The experiments witness to the remarkable skill of the surgeon and the extraordinary capacity of the bovine embryo to recover from heroic surgery which literally cleaves the embryo and its membranes into two (Figure 5-9). Several sets of monozygotic twins have been produced (Figure 5-10) by this impressive microsurgery (Ozil, 1983).

It would seem that if embryo bisection in mice, sheep, and cattle results in experimental twinning, then the procedure would similarly work with horses. It does. Twin foals have been reported (Allen, 1980).

The significance of the technically difficult embryo bisections

93

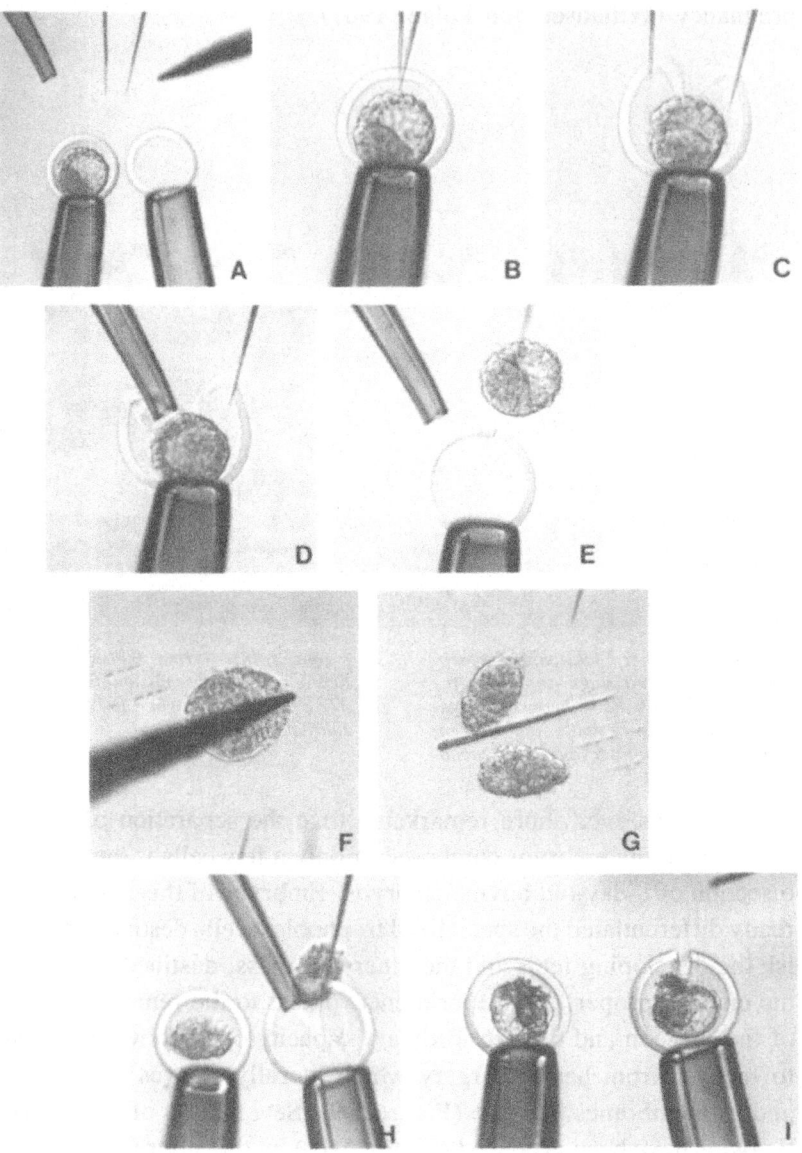

FIGURE 5-9. *Surgical bisection of an 8-day bovine embryo. (A) Embryo in its zona pellucida held by a pipette. The inner cell mass is clearly visible on the lower left of the embryo; the empty zona pellucida is on the right. (B & C) Sharp microneedles cut and open the zona pellucida. (D) Fluid is introduced via a micropipette, upper left, which (E) expels the embryo. (F & G) The living embryo is bisected with a sharp microscalpel. (H & I) Each half embryo is inserted into a zona pellucida and subsequently incubated, after which each will be transferred to a recipient cow. (From J. P. Ozil, "Production of identical twins by bisection of blastocysts in the cow." Journal of Reproduction and Fertility, 69:463-68)*

FIGURE 5-10. *Two sets of mono-zygotic twins produced by transfer of bisected embryos. (From Willadsen, 1981. Reproduced by permission from* Nature, *277:298-300, copyright © 1979 Macmillan Journals Limited)*

with subsequent survival of the experimental animals suggests that it is probably only a matter of time before the somewhat more complex procedure of nuclear transplantation will be successful in mammals other than mice (Seidel, 1983). The economic benefits of cloning large domestic animals is already being discussed (Van Vleck, 1981).

HUMAN CLONING: THE HOW TO

A skeptic might argue that it is premature to outline a procedure for human cloning. I have already asserted that no human has been cloned and will, in the Epilogue, give the reasons why I believe that human cloning will be an extraordinarily unlikely event. However that may be, the fact remains that the procedures that make human cloning a real possibility are at hand.* Further, a knowledge of the biological possibilities is a better basis for judgment than is groping in the dark. I hope the following provides a glimmer of light.

What is required? A means of obtaining a sufficient number of ova so that experimental procedures may be performed; an enucleation procedure; cells to provide nuclei for transplantation; a nuclear-

*I am not unique in considering the possibility of human cloning; an editorial writer for the British journal *Nature* wrote: "If ethical committees wish to brood about *something tangible*, they should worry about cloning. Nature 295:445 (1982). (italics added)

insertion procedure; an *in vitro* culture technique; and embryo transfer to a foster mother. All these methodologies are available.

Before going into detail on how to clone a human, let me note that there is at least one principal difference between mouse and human cloning (aside from the obvious one that humans are different from lower animals). Mouse nuclear transplantation, thus far, has used enucleated eggs that were fertilized by natural mating before enucleation. Hardly anyone thinks ill of the experimenter who snips out the oviduct of a mouse to recover fertilized ova. Human fertilization also occurs in the oviduct. But few women would consent to having their tubes removed in order to assist in human cloning. As an alternative to the mouse procedure of zygote recovery from the oviduct, *in vitro* fertilization would probably be the method of choice. These procedures if applied to humans raise *horrendous* ethical issues, which I will discuss in the Epilogue.

Ovulation and Egg Recovery

Ovulation in humans occurs about midway in the menstrual cycle. Day one of a typical 28-day cycle is the time of bleeding onset—the start of menstruation. Ovulation occurs about 2 weeks later. During these 14 days, a previously inactive immature egg (oocyte) starts to mature. It is enclosed in an ovarian follicle. The preovulatory follicle grows rapidly until it is no longer microscopic. By the time of ovulation, it is a half inch or more in diameter and takes the form of an elevated bump on the surface of the ovary. The bump has been compared with a blister because the preovulatory follicle contains fluid. Within the blister-like follicle is a small mound of cells which contains the growing oocyte. That mound of cells will remain attached to the ovum after ovulation. It is these ovarian follicle cells which, as in the mouse, are known as cumulus cells. The oocyte at ovulation moves directly to the upper part of the oviduct (Fallopian tube). It would be too late to recover an egg after it is ovulated because it would be virtually impossible to retrieve that speck of protoplasm deep within the human female. The time to obtain the ovum would be before release from the ovary. The ovary is relatively large (about 1 ½ inches × 1 inch × ⅓ inch), and, as has been stated, the follicle that contains the oocyte is also large (½ inch or larger in diameter [Lemay et al., 1982]). Thus, it would be easier to obtain an oocyte from the ovary than from any other place in the reproductive tract.

There is an instrument that permits a visual probe of the interior of a woman's body called a laparoscope (Feichtinger et al., 1981). In theory, at least, one could probe the interior of a potential egg donor daily as the time of ovulation approaches to ascertain if ovulation is near. In as much as use of the laparoscope involves a small incision through the abdominal wall accompanied by inflation of the abdomen with carbon dioxide or a mixture of other gases (inflation is called pneumoperitoneum) and general anesthesia (Wood et al., 1981), it is unlikely that many women would particularly enjoy serial observation of their ovaries. Consequently, it would be useful if there were a non-invasive procedure for predicting the precise time of ovulation. There is.

Luteinizing hormone is produced by the anterior pituitary gland. The hormone is liberated in increasing quantities just before natural ovulation. The increase is known as the luteinizing hormone *surge*. The level of the hormone can be monitored with urine or blood samples. Further, the growth of the follicle can be followed by visualization with ovarian ultrasonography (de Crespigny et al., 1981). Ultrasound diagnosis has the capability of revealing on which side the ripening follicle will be found, and it eliminates the necessity of laparoscopic examination in those women who have already ovulated. Because the diameter of the follicle can be measured with ultrasound, this procedure and the luteinizing hormone surge indicate the time that the follicle is ripe for the retrieval of the oocyte (Trounson and Conti, 1982; Lemay et al., 1982; Leeventveld et al., 1983).

When the time is appropriate a hollow aspiration needle is inserted into one or more ripe follicles under direct visual guidance of the laparoscope. The oocyte is removed with some follicular fluid (Jones et al., 1982). It is not a hit-or-miss procedure—experienced laparoscopists have recovery rates in excess of 90%. Although the yield of ripe follicles is usually one, the number can be increased by stimulation. Clomiphene citrate (Trounson et al., 1981), or this drug used in combination with other drugs (Trounson, 1982; Leeventveld et al., 1983; Laufer at al., 1983) has been used to enhance the number of eggs that can be retrieved at any one time.

Enucleation

The oocytes obtained from ripe ovarian follicles are, of course, not fertilized. The two laboratories that have reported success with mouse cloning enucleated *fertilized* ova. Because human eggs resemble mouse eggs far more than they do frog eggs, I predict that human cloners would follow the mouse model. Hence, before enucleation, the ovum would be fertilized.

Fertilization *in vitro* of human ova may have had its first success about 40 years ago (Rock and Menkin, 1944). These early investigators examined nearly 800 human eggs and they reported cleavage of 3 which had been inseminated. The paper is interesting not only because of *the early human fertilization* under laboratory conditions but because of their citation of *in vitro* fertilization experiments in small mammals dating back to the late nineteenth century. Their pioneering fertilization study was described in greater detail in a later publication (Menkin and Rock, 1948).

Many recent studies have detailed the procedures for successful *in vitro* human fertilization (Edwards et al., 1969; Steptoe and Edwards, 1978; Nayudu et al., 1983; Fishel et al., 1984). Briefly, the oocytes harvested from the ovary are incubated in a culture medium for several hours to permit maturation (recall that they are plucked from the ovary a brief time before natural ovulation; thus, they are not as mature as spontaneously ovulated ova). Semen obtained from a donor several hours before the desired time of fertilization is washed and centrifuged (Edwards et al., 1970). The sperm are suspended in the fertilization culture medium and diluted to the proper concentration for fertilization. Unwashed and undiluted sperm will not fertilize ova *in vitro*. Since this sort of fertilization occurs outside the body, it is sometimes referred to as "extracorporeal fertilization" or "external human fertilization" (Grobstein, 1981).

Cells of many types, including sperm, can be frozen (Foote, 1981; Phillip et al., 1983). Because of this, the donor's semen may be stored frozen and used for insemination after thawing. Thawed semen is about as effective as fresh semen in fertilization.

Sperm penetrate the human ovum within 3 to 6 hours after insemination, and both pronuclei are readily identifiable by 12 hours (Trounson and Conti, 1982). It is at this time that enucleation is feasible. Probably nothing need be added to the description I have given of mouse-egg enucleation. The ovum could be enucleated surgically be penetration with a micropipette, or it could be enucleated by aspiration of a bleb of cytoplasm containing both the male and female pronuclei.

Cells for Nuclear Transplantation

Thus far, after a third of a century, only embryonic nuclei have proved competent to program for total development in the frog when transplanted to an enucleated egg (Chapter 3). The scant experimental data available suggest similar results with mice. If this proves to be true with additional mouse experiments, then it is likely that only young embryonic *human* nuclei would be competent to program for normal development with the cloning procedure. Thus, the experimenter with human material would seek nuclei from the inner-cell mass of the early human embryo. The procedure for obtaining inner-cell-mass nuclei should be no different from that for the mouse. The zona pellucida must be removed from a cultured embryo, the trophectoderm separated from the inner-cell mass, and, finally, the cells dissociated with an appropriate enzyme in a calcium- and magnesium-free salt solution.

Transplantation of Human Inner-Cell-Mass Nuclei

Obviously, no one at this juncture knows the optimal method for joining a donor nucleus (from the inner-cell mass) with enucleated cytoplasm (from an *in vitro* fertilized ovum). However, reference to already existing mouse work suggests that surgical implantation with a micropipette, or fusion with an inactivated Sendai virus, would provide a viable clone.

The human nuclear transplant then needs to be cultured, as already described for *in vitro* fertilization, until it is ready for implantation into a human foster mother. Embryo transfer would likely be attempted when the clone has reached the 8- to 16-cell stage. Embryo transfer could be delayed, if desired, by freezing the clone until needed (Schneider and Maurer, 1983).

Human Embryo Transfer

The clone would be drawn into a fine plastic tube (a catheter) which, in turn, would be introduced through the cervical canal into the interior of the uterus. The procedure would be accomplished without anesthesia (Wood et al., 1981; Trounson and Conti, 1982; Leeton et al., 1982). Human embryo-transfer technology is changing and, therefore, new procedures will likely evolve in the next few years. There is little to be gained by a comparison of techniques, types of tubes for embryo insertion, position of the would-be foster mother, etc. However, it can be stated that multiple embryo transfer seems to enhance the pregnancy rate, and it may be that if ever human cloning is attempted, the experimenter would introduce more than one clone per operation. The foster mother and experimenter would then have a wait of about 9 months for the outcome.

As has been stated throughout this book, no human has been cloned. I doubt if one will be. Nevertheless, if we are to deal with the issue rationally, an exposition on biological possibilities is essential.

 **Epilogue: An Essay on Human Cloning**

I believe human cloning is an inappropriate endeavor for biological or medical scientists, primarily for economic and ethical reasons.

The word that best describes funds available for biomedical research, from both government agencies and private organizations, is "inadequate." I think it is unlikely that there will *ever* be funds available to provide economic support for all potential research projects. Consequently, judgments must be made concerning what studies are to be funded, and priorities must be established so that the limitations imposed by economic considerations do not impede research essential to the welfare of humankind. Is research that might increase the number of human genetic entities (i.e., research that would lead to the production of cloned human replicates) more important than research that seeks to improve the quality of life for humans already in existence? My answer is an unequivocal "No."

We have pressing problems, the solution of which will enhance the lives of many. We need clean air, good food, and pure water—none of which is likely to happen without extensive ecological research combined with technology developed by civil and chemical engineers. I think this research deserves a high priority for funding. There are special maladies of the mature and ills of the aged. Our attention to these concerns is a measure of the quality of our civilization. Human relationships, and how they can be made more compassionate, is something we need to learn much about.

There are, then, many areas of research that have an intimate impact on the welfare of humankind. I assert as vigorously as I can that funding for research that will enhance the lives of humans already here should have priority over funding for research into human cloning—which, of course, it already does.

To clone a mouse means, with present knowledge and technology, to insert a body-cell nucleus into an enucleated fertilized egg (Chapter 5). I presume that a similar procedure must be followed to clone a human. And, although mouse cloning *is* ethical, I have profound reservations about whether it will ever be ethical to clone humans. I believe it is unethical because of the procedures required to obtain and manipulate oocytes, and because the transfer of the clone to the uterus of the foster mother-to-be is not without hazard.

To conceive through conventional sexual intercourse is to expose both baby and mother to certain dangers that are well known and need not be documented here. There seems to be only a small additional hazard when the conception of the baby occurs *in vitro* instead of in the Fallopian tubes (Grobstein et al., 1983). That extra hazard, perhaps associated with laparoscopy for oocyte recovery, is obviously acceptable to those couples who would be unable to conceive without the procedure. Although considerable attention has been lavished on human fertilization outside the body, it is, nevertheless, simply a modification of what occurs naturally. Humans, conceived naturally, result from the union of a sperm with an ovulated oocyte. Humans, conceived by *in vitro* fertilization, result from the union of a sperm with a surgically plucked oocyte. The essential difference is the site of fertilization—internal, in the Fallopian tube, or external, in laboratory glassware. The cellular components, the genetic significance, and the end results are the same.

Human cloning would be different. Initially, there would be a period of experimentation. By analogy with mouse cloning (Chapter 5), fertile eggs would have to be obtained. Although hazard to the human oocyte donor would be very small, it would be a hazard imposed on a healthy woman to *neither prevent nor cure a disease*. The hormones used to induce the multiple ovulation necessary for the procedure would be drugs not to eliminate or control an illness, but drugs administered to enable experimentation. The surgical incisions to admit the laparoscope would be made not to remove diseased tissue or repair internal organs but to recover oocytes for experimentation. I find it repugnant to administer drugs to or perform surgery on healthy humans for nonmedical procedures. The process becomes more odious when we surgically manipulate a fertilized egg.

A sperm is alive but not usually considered an organism. The same is true of an egg. But a human egg fertilized with a human sperm is a nascent being. Whether or not that individual has the same legal rights as an individual after birth is not a question to be debated here. What I wish to note is that it is an immature, developing being. To clone a human, the fertilized egg must be enucleated. That means that a nascent being must be eliminated, and the elimination procedure, at least during the developmental phase of human cloning research, would have no significance other than experimental.

There is no clone without nuclear insertion. Whether inserted surgically, or fused to the egg via an inactivated virus, a nuclear donor must be found. The nuclear donor, of necessity with current technology, would be embryonic. Another nascent being must be destroyed to provide dissociated cells for the nuclear donor.

Embryo transfer would follow after a successful period of culture *in vitro*. Embryo transfer in cattle is a thriving business (Seidel, 1981). Embryo transfer in humans is a procedure still evolving (Wood et al., 1981; Trounson and Conti, 1982; Leeton et al., 1982; Nayudu et al., 1983). It can be anticipated that occasionally, just as in normal pregnancy, an ectopic implantation would occur. "Ectopic" simply refers to an inappropriate site of embryo implantation—a site such as a Fallopian tube. Thus, there would be a modest hazard to the woman bearing an ectopic clone. For what purpose?

I can think of few biological experiments with human cloning that could not be better made with frog or mouse material. There would, of course, be new information about the relative roles of genetics and environment, because instead of a limited number of identical twins who have been reared apart, one could devise experiments with 10 or 20 clones—each reared under different environmental conditions. New information, it is true, but were the experiments to be performed, it would be information gained at the expense of gamete manipulation coupled with infant and child abuse. How else would one measure the effect of different environments on a clone (in the aggregate) without rearing the isogenic group apart? Is "abuse" an adequate way to characterize the behavior of an individual or committee that would by design separate such closely related individuals for a rearing experiment?

The production of human clones is the material of science fiction—the scientists that I know would not engage in such endeavors.

The reader may well agree that ethical scientists have no intention of cloning humans. But, as a friend of mine said, "Minnesota is knee-deep in frogs, but frogs are cloned at the University." One may disagree with the assertion that cloning of humans is unlikely because there is no need to clone or because it is unethical. Some skeptical people may believe that cloning will occur because cloning is possible and because someone somewhere may want to clone. Since history reveals *Homo sapiens* to be perverse from time to time, how would consideration of the ethics of gamete manipulation or the burgeoning global population deter human cloning if an unethical person chooses to clone? My answer to that, quite frankly, is that just as an occasional individual descends to the illegal and immoral (humans engage in counterfeiting money, robbing storekeepers, shooting neighbors, and other assorted mayhem), probably someday, somewhere, an ill-advised scientist will successfully clone a human. Will that act endanger our species? I think not.

It is my judgment that human cloning, should it ever occur, will be an isolated phenomenon with little if any lasting significance. No armies will be produced. No clones of cosmonauts will emerge. Perhaps because of the misled quest of a curious individual, one or more cloned boys or girls will appear. They will mature but will have no special attributes other than genetic identity with the nuclear donor. They will not be a hazard to others. The cloned people will be people no more distinguishable from their fellows than cloned frogs are from frogs that breed naturally in the Minnesota outback.

Dr. C. Everett Koop, the U.S. Surgeon General, in his 1982 report on health and smoking, identified cigarette smoking as "the chief, single avoidable cause of death in our society and the most important health issue of our time." Clearly, tobacco constitutes an enormous hazard. There are other hazards. Armies with conventional weapons can be moved on short notice, and these weapons have an awesome destructive capacity. Nuclear weapons possess so much destructive power that it is difficult to imagine civilization surviving a conflict in which they are used. Real hazards plague modern humans. The hazard of human cloning is fantasy compared with these.

I do not wish to trivialize the apprehensiveness of those who fear cloning. However, let me repeat again that cloned humans will be no more different from ordinary humans than a frog produced by cloning is different from a frog spawned in a northern pond. The skeptic reading that sentence could argue that the fear of thoughtful people is not the cloning of an occasional man or woman. Their fear, they might say, concerns clones in the aggregate. Hoards of soldiers—or hoards of workers—produced at the laboratory bench. The argument might state that a particular genetic constitution would result in soldiers of extraordinary merit because of courage, aggressiveness, and endurance. Factory assembly-line workers might be cloned who would not tire of repetitiveness and who would possess adaptability and judgment far greater than the best of robots. I believe that concern about the cloning of hoards of humans is naive for two reasons, economic and biological.

I remember my early years as a postdoctoral associate at a cancer research institute in Philadelphia. I was appalled at the cost of maintaining the mouse colony. That was 25 years ago. I am still impressed with the cost of maintaining animals for research. It is estimated that my university spends about 3.2 million dollars per year on animal care. I can assure readers that mouse husbandry, as costly as it is, is far cheaper than any mode of human husbandry.

How much would it cost for a government, or a giant multinational corporation, to rear a colony of human clones? Cost estimates might be made by comparison to prison care, hospital care, or college tuition with room and board included. Concerning the latter, the college would have to be private because tuition of state-supported institutions is highly subsidized. Estimates probably would vary from a minimum of $10,000 per cloned human per year to perhaps $100,000—or even more. The figure would have to be multiplied by 18 or 21 (with a factor for inflation) to arrive at an estimate for the cost of rearing the cloned soldier (or cloned assembly-line worker) because soldiers and workers do not perform their assigned tasks until they have the physical maturity to do so.

With a Minnesota outback "knee-deep in frogs," one does not clone to produce more frogs. One clones to increase understanding of cancer, aging, and other challenges. With a number approaching 5 billion people on earth, there is no need to manufacture more. Scrutiny of the unemployment rate of most developed countries reveals *plenty* of bodies available for armies or industries. The construction of fabricated humans in quantity for any purpose is an unreasonable proposition if for no other reason than economics.

However, biology provides another reason why clones would not be fabricated in the aggregate. We do not know what biological traits would be best for future tasks. It is difficult to imagine knowing what kind of army (air force, navy) or industrial establishment we will have in the twenty-first century. How does a society or committee or autocrat select a genotype in the 1980s for a task two decades or so into the future?

Further, it is false to assume that there is *one* best genetic constitution for a particular task. Diversity is essential for any vocation that can be imagined. Consider a college or university. The genetic constitution of a particularly successful physicist might be thought worthy of amplification. Would that genotype serve well as an art historian, law professor, or botanist? Or, for that matter, would the physics department of the hypothetical professor desire to be peopled with only physicists of that particular type and personality? Could you get the physics department to agree that the genetic constitution of *any* individual is what they want at the outset of the next century? The assumptions relating to the concept of a "best genetic constitution" are only part of the problem. The rest of the problem relates to the fact that it would be impossible ever to recreate personality. Perhaps recreation of the soma will be possible, but personality is a far more complex matter. The hypothetical physics professor would have been a product of his time and his unique experiences, coupled with his genetic constitution. His environment and times could never be cloned even if his body could be. His parents, his siblings, his schoolmates, his teachers, and all of his life experiences would have to be recreated to recreate him. That clearly is not possible.

There is speculation about the famous and the infamous. What would the clone of Adolph Hitler, Albert Einstein, or John Fitzgerald Kennedy be like? Why not Eleanor Roosevelt, Golda Meir, Mahatma Gandhi, Martin Luther King, or Rachel Carson clones? In my opinion all such speculation is unproductive. Einstein and Meir and the others were never known as embryos. These names have meaning only when associated with their extraordinarily notable adult lives. Thus, speculation about clones of adults has meaning *only* in relationship to the capacity to clone adults, which is not possible with present technology. A third of a century of amphibian cloning experiments (Chapters 3 and 4) has firmly established the totipotency of early embryonic nuclei and the *lack* of totipotency of nuclei obtained from adult organisms. Although we cannot know the results of future mouse-cloning experiments, it would seem likely that they will parallel the frog-cloning experiments (Chapter 5). Consequently, it is unlikely that human clones will be produced from adult nuclear donors any time in the near future (if ever); thus, speculation is only speculation.

I have written this chapter because fear about human cloning detracts from the potential benefits that cloning research may provide. Cloning studies of frogs have provided insight into the role of the genetic material DNA in developing creatures. If cloning research provided no other justification for its being, added insight into and understanding of developmental problems would be enough. However, some cloning studies relate to specific problems that affect the well being and survival of humankind. It is that fact, the potential for human good of cloning research, that I wish to leave with you.

References

Ackerman SB, RJ Swanson, GK Stokes, LL Veeck. 1984. Culture of mouse preimplantation embryos as a quality control assay for human *in vitro* fertilization. Gamete Res 9:145–152.

Allen WR. 1980. Producing twin foals. Vet Rec 107:50.

Backs-Hüsemann, D, J Reinert. 1970. Embryo formation by isolated single cells from tissue cultures of *Daucus carota*. Protoplasma 70:49–60.

Baranska W, H Koprowski. 1970. Fusion of unfertilized mouse eggs with somatic cells. J Exp Zool 174:1–14.

Barber MA. 1911. A technique for the inoculation of bacteria and other substances into living cells. J Infect Dis 8:348–360.

Barski G, S Sorieul, F Cornefert. 1961. "Hybrid" type cells in combined cultures of two different mammalian strains. J Nat Cancer Inst 26:1269–1291.

Biggers JD, WK Whitten, DG Whittingham. 1971. The culture of mouse embryos *in vitro*. In: JC Daniel Jr (ed) Methods in Mammalian Embryology. San Francisco, Freeman, pp 86–116.

Binns AN, HN Wood, AC Braun. 1981. Suppression of the tumorous state in crown gall teratomas of tobacco: A clonal analysis. Differentiation 19:97–102.

Brachet A. 1912. Developpement *in vitro* de blastomeres et jeune embryons de mammiferes. C R Hebd Seances Acad Sci 155:1191.

Brack C, M Hirama, R Lentiard-Schuller, S Tonegawa. 1978. A complete immunoglobulin gene is created by somatic recombination. Cell 15:1–14.

Braun AC. 1965. The reversal of tumor growth. Sci Am 213(5):75–83.

Braun AC, HN Wood. 1976. Suppression of the neoplastic state with the acquisition of specialized functions in cells, tissues, and organs of crown gall teratomas of tobacco. Proc Nat Acad Sci USA 73:496–500.

Briggs R, EU Green, TJ King. 1951. An investigation of the capacity for cleavage and differentiation in *Rana pipiens* eggs lacking "functional" chromosomes. J Exp Zool 116:455–500.

Briggs R, TJ King. 1952. Transplantation of living nuclei from blastula cells into enucleated frogs' eggs. Proc Nat Acad Sci USA 38:455–463.

Briggs, R, TJ King. 1957. Changes in the nuclei of differentiating endoderm cells as revealed by nuclear transplantation. J Morphol 110:269–311.

Brinster RL. 1971. Measuring embryonic enzyme activity. In: JC Daniel Jr (ed) Methods in Mammalian Embryology. San Francisco, Freeman, pp 215–227.

Broad WJ. 1981. Saga of Boy Clone Ruled a Hoax. Science 211:902.

Bromhall JD. 1975. Nuclear transplantation in the rabbit egg. Nature 258:719–722.

Brun RB. 1978. Developmental capacities of *Xenopus* eggs, provided with erythrocyte or erythroblast nuclei from adults. Dev Biol 65:271–284.

Caplan AL, Herrera-Estrella, D Inzé, E Van Haute, M Van Montagu, J Schell, P Zambryski. 1983. Introduction of genetic material into plant cells. Science 222:815–821.

Chabry L. 1887. Contribution à l'embryologie normale et tétratologique des ascidies simples. J Anat Physiol Norm Pathol Homme Anim 23:167–319.

Chaleff RS. 1983. Isolation of agronomically useful mutants from plant cell cultures. Science 219:676–682.

Chambers R, EL Chambers. 1961. Explorations into the nature of the living cell. Cambridge, Mass, Harvard University Press.

Chen LT, YC Hsu. 1982. Development of mouse embryos *in vitro*: Preimplantation to the limb bud stage. Science 218:66–68.

Cole CJ. 1984. Unisexual lizards. Sci Am 250(1):94–100.

Cook RE. 1983. Clonal plant populations. Am Sci 71:244–253.

Daniel JC Jr, K Takahashi. 1965. Selective laser destruction of rabbit blastomeres and continued cleavage of survivors *in vitro*. Exp Cell Res 39:475–482.

De Cherney AH. 1983. Doctored babies. Fertil Steril 40:724–727.

de Condolle A. 1886. Origin of Cultivated Plants. 2nd Edition. Reprinted 1959. New York, Hafner Publishing Co.

de Crespigny LJCh, C O'Herlihy, IJ Hoult, HP Robinson. 1981. Ultrasound in an *in vitro* fertilization program. Fertil Steril 35:25–28.

de Fonbrune P. 1949. Technique de micromanipulation. Paris, Masson et Cie.

Di Berardino MA. 1979. Nuclear and chromosomal behavior in amphibian nuclear transplants. Int Rev Cytol, Suppl 9:129–160.

Di Berardino MA. 1980. Genetic stability and modulation of metazoan nuclei transplanted into eggs and oocytes. Differentiation 17:17–30.

Di Berardino MA, NJ Hoffner. 1970. Origin of chromosomal abnormalities in nuclear transplants—a reevaluation of nuclear differentiation and nuclear equivalence in amphibians. Dev Biol 23:185–209.

Di Berardino MA, NJ Hoffner. 1971. Development and chromosomal constitution of nuclear transplants derived from male germ cells. J Exp Zool 176:61–72.

Di Berardino MA, NJ Hoffner. 1983. Gene reactivation in erythrocytes; nuclear transplantation in oocytes and eggs of Rana. Science 219:862–864.

Di Berardino MA, NJ Hoffner, L Etkin. 1984. Activation of dormant genes in specialized cells. Science. 224:946–952.

Di Berardino MA, TJ King, RG McKinnell. 1963. Chromosome studies of a frog renal adenocarcinoma line carried by serial intraocular transplantation. J Nat Cancer Inst 31:769–789.

Di Berardino MA, M Mizell, NJ Hoffner, DG Friesendorf. 1983. Frog larvae cloned from nuclei of pronephric adenocarcinoma. Differentiation 23:213–217.

Dickmann Z. 1971. Egg transfer. In: JC Daniel Jr (ed) Methods in Mammalian Embryology. San Francisco, Freeman, pp 133–145.

Dodds JH, LW Roberts. 1982. Experiments in Plant Tissue Culture. Cambridge, Cambridge University Press.

Driesch H. 1892. Entwicklungsmechanische Studien. I. Der Wert der beiden ersten Furchungszellen in der Echinodermenentwicklung. Experimentelle Erzeugung von Teil und Doppel bildungen. Z Wiss Zool 53:160–178, 183–184. English translation by L Mezger, M and V Hamburger, TS Hall. In: BH Willier, JM Oppenheimer (eds) 1964. Foundations of Experimental Embryology, Englewood Cliffs, New Jersey, Prentice-Hall, pp 38–50.

Drost M, JM Wright Jr, WS Cripe, AR Richter. 1983. Embryo transfer in water buffalo (*Bubalus bubalis*). Theriogenology 20:579–584.

Du Pasquier L, MR Wabl. 1977. Transplantation of nuclei from lymphocytes of adult frogs into enucleated eggs: Special focus on technical parameters. Differentiation 8:9–19.

Edwards RG, BD Bavister, PC Steptoe. 1969. Early stages of fertilization *in vitro* of human oocyte matured *in vitro*. Nature 221:632–635.

Edwards RG, PC Steptoe, JM Purdy. 1970. Fertilization and cleavage *in vitro* of preovular human oocytes. Nature 227:1307–1309.

Ellinger MS, DR King, RG McKinnell. 1975. Androgenetic haploid development produced by ruby laser irradiation of anuran ova. Radiat Res 62:117–122.

Elsdale TR, M Fischberg, S Smith. 1958. A mutation that reduces nucleolar number in *Xenopus laevis*. Exp Cell Res 14:642–643.

Elsdale TR, JB Gurdon, M Fischberg. 1960. A description of the technique for nuclear transplantation in *Xenopus laevis*. J Embryol Exp Morphol 8:437–444.

Fedoroff NV. 1984. Transposable genetic elements in maize. Sci Am 250(6):84–98.

Feichtinger W, S Szalay, A Beck, P Kemeter, H Janisch. 1981. Results of laparoscopic recovery of preovular human oocytes from nonstimulated ovaries in an ongoing *in vitro* fertilization program. Fertil Steril 36:707–711.

Fishel SB, RG Edwards, JM Purdy. 1984. Births after a prolonged delay between oocyte recovery and fertilization in vitro. Gamete Res 9:175–181.

Fletcher J. 1974. The Ethics of Genetic Control: Ending Reproductive Roulette. Garden City, New York, Anchor Press/Doubleday.

Foote RH. 1981. The artifical insemination industry. In: BG Brackett, GE Seidel, SM Seidel (eds) New Technologies in Animal Breeding. New York, Academic Press, pp 13–39.

Fowler RE, RG Edwards. 1957. Induction of superovulation and pregnancy in mature mice by gonadotropins. J Endocrinol 15:374–384.

Franklin D. 1984. Embryo transfer: It's a boy. Science News 125:85

Frost HB. 1938. Nucellar embryony and juvenile characters in clonal varieties of *Citrus*. J Hered 29:423–432.

Gasaryan KG, NM Hung, AA Neyfakh, VV Ivanenko. 1979. Nuclear transplantation in teleost *Misgurnus fossilis* L. Nature 280:485–487.

Gates AH. 1971. Maximizing yield and developmental uniformity of eggs. In: JC Daniel Jr (ed) Methods in Mammalian Embryology. San Francisco, Freeman, pp 64–75.

Gilbert W, L Villa-Komaroff. 1980. Useful proteins from recombinant bacteria. Sci Am 242(4):74–94.

Goldman MA, GP Holmquist, MC Gray, LA Caston, A Nag. 1984. Replication timing of genes and middle repetitive sequences. Science 224:686–692.

Graham CF. 1971. Virus assisted fusion of embryonic cells. Acta Endocrinol, Suppl 153:154–167.

Greenhouse, DD. 1984. Holders of inbred and mutant mice in the United States. ILAR News 27: 1A-30A.

Grobstein C. 1981. From Chance to Purpose: An Appraisal of External Human Fertilization. Reading, Massachusetts, Addison-Wesley Publishing Company.

Grobstein C, M Flower, J Mendeloff. 1983. External human fertilization: An evaluation of policy. Science 222:127–133.

Gurdon, JB, S Brennan, S Fairman, TJ Mchun. 1984. Transcription of muscle-specific actin genes in early Xenopus development: Nuclear transplantation and cell dissociation. Cell 38:691–700.

Gurdon JB, RA Laskey, OR Reeves. 1975. The developmental capacity of nuclei transplanted from keratinized cells of adult frogs. J Embryol Exp Morphol 34:93–112.

Gustafsson A. 1946. Apomixis in higher plants. Part I. The mechanism of apomixis. Lunds Universitets Arsskr, NF Bd 42:1–66.

Haccius B. 1978. Question of unicellular origin of non-zygotic embryos in callus cultures. Phytomorphology 28:74–81.

Harvey EB. 1940. A comparison of the development of nucleate and nonnucleate eggs of *Arbacia punctulata*. Biol Bull (Woods Hole, Mass) 79:166–187.

Hayflick L. 1977. The cellular basis for biological aging. In: CE Finch, L Hayflick (eds) Handbook of the Biology of Aging. New York, Van Nostrand Reinhold, pp 159–186.

Heape W. 1890. Preliminary note on the transplantation and growth of mammalian ova within a uterine foster-mother. Proc R Soc London 48:457–458.

Hearing before the Subcommittee on Health and the Environment. Developments in Cell Biology and Genetics. Serial No 95–105. Washington, DC, US Government Printing Office, 1978.

Hoppe PC, K Illmensee. 1977. Microsurgically produced homozygous-diploid uniparental mice. Proc Nat Acad Sci USA 74:5657–5661.

Hoppe PC, K Illmensee. 1982. Full-term development after transplantation of parthenogenetic embryonic nuclei into fertilized mouse eggs. Proc Nat Acad Sci USA 79:1912–1916.

Hopwood DA. 1981. The genetic programming of industrial microorganisms. Sci Am 245(3):67–78.

Horsch RB, RT Fraley, SG Rogers, PR Sanders, A Lloyd, N Hoffmann. 1984. Inheritance of functional foreign genes in plants. Science 223:496–498.

Illmensee K. 1973. The potentialities of transplanted early gastrula nuclei of *Drosophilia melanogaster*. Production of the imago descendents by germline transplantation. Wilhelm Roux Arch Entwicklungsmech Org 171:331–343.

Illmensee K. 1984. Experimental genetics of the mammalian embryo. Pontif Acad Sci Scr Varia 51:165–179.

Illmensee K, PC Hoppe. 1981. Nuclear transplantation in *Mus musculus*. Developmental potential of nuclei from preimplantation embryos. Cell 23:9–18.

Johnson IS. 1983. Human insulin from recombinant DNA technology. Science 219:632–637.

Johnson MH. 1981. The molecular and cellular basis of preimplantation mouse development. Biol Rev 56:463–498.

Jones HW Jr, AA Acosta, J Garcia. 1982. A technique for the aspiration of oocytes from human ovarian follicles. Fertil Steril 37:26–29.

Kemperman JA, BV Barnes. 1976. Clone size in American aspens. Can J Bot 54:2603–2607.

King TJ, R Briggs. 1953. The transplantability of nuclei of arrested hybrid blastulae (*Rana pipiens* ♀ x *Rana catesbeiana* ♂). J Exp Zool 123:61-78.

King TJ, R Briggs. 1954. Transplantation of living nuclei of late gastrulae into enucleated eggs of *Rana pipiens*. J Embryol Exp Morphol 2:73–80.

King TJ, R Briggs. 1955. Changes in the nuclei of differentiating gastrula cells, as demonstrated by nuclear transplantation. Proc Nat Acad Sci USA 41:321–325.

King TJ, RG McKinnell. 1960. An attempt to determine the developmental potentialities of the cancer cell nucleus by means of transplantation. In: Cell Physiology of Neoplasia. Austin, University of Texas Press, pp 591–617.

Konar RN, E Thomas, HE Street. 1972. Origins and structure of embryoids arising from epidermal cells of the stem of *Ranunculus sceleratus*. J Cell Sci 11:77–93.

Kraemer DC. 1983. Intra- and interspecific embryo transfer. J. Exp Zool 228:363–371.

Lambeth VA, CR Leoney, SA Voelkel, DA Jackson, KG Hill, RA Godke. 1983. Microsurgery on bovine embryos at the morula stage to produce monozygotic twin calves. Theriogenology 20:85–96.

Laufer N, AH De Cherney, FP Haseltine, ML Polan, HC Mezer, AM Dlugi, D Sweeney, F Nero, F Naftolin. 1983. The use of high-dose human menopausal gonadotropin in an *in vitro* fertilization program. Fertil Steril 40:734–741.

Lederberg J. 1972. Biomedical frontiers: genetics. In: RM Kunz, H Fehr (eds) The Challenge of Life. Basel and Stuttgart, Birkhäuser Verlag, pp 231–245.

Leeventveld RA, I Van Gent, ATh Alberda, JW Wladimiroff. 1983. Assessment of follicular development in clomiphene induced cycles by means of ultrasound and laparoscopy: A comparative study. Ultrasound Med Biol 9:595–598.

Leeton J, A Trounson, D Jessup, C Wood. 1982. The technique for human embryo transfer. Fertil Steril 38:156–161.

Lejeune J, M Gautier, R Turpin. 1959. Etude des chromosomes somatique de neuf enfants mongoliens. C R Hebd Seances Acad Sci 248:1721–1722.

Lemay A, A Bastide, R Lambert, JE Rioux. 1982. Prediction of human ovulation by rapid luteinizing hormone (LH) radioimmunoassay and ovarian ultrasonography. Fertil Steril 38:194–201.

Lin TP, JY Chan. 1981. Laser microbeam inactivation of mouse egg nucleus. J Cell Biol 91:(2 part 2)170a.

Lin TP, J Florence, JO Oh. 1973. Cell fusion induced by a virus within the zona pellucida of mouse eggs. Nature 242:47–49.

Loeb J. 1894. Uber eine einfache Methode, zwei order mehr zusammengewachsene Embryonen aus einem Ei hervorzubringen. Pfluegers Arch 55:525–530.

Luchtel D, JG Bluemink, SW de Laat. 1976. The effect of injected cytochalasin B on filament organization in the cleaving egg of *Xenopus laevis*. J Ultrastruct Res 54:406–419.

Lucké B. 1934a. A neoplastic disease of the kidney of the frog, *Rana pipiens*. Am J Cancer 20:352–379.

Lucké B. 1934b. A neoplastic disease of the kidney of the frog, *Rana pipiens*. II. On the occurrence of metastasis. Am J Cancer 22:326–334.

Lucké B, H Schlumberger. 1949. Induction of metastasis of frog carcinomas by increase of environmental temperature. J Exp Med 89:269–278.

Magnuson T, CJ Epstein. 1981. Genetic control of very early mammalian development. Biol Rev 56:369–408.

Maniatis T, EF Fritsch, J Sambrook. 1982. Molecular cloning. A Laboratory Manual. Cold Spring Harbor, New York, Cold Spring Harbor Laboratory.

Mareel MMK, M DeBrabander. 1978. Effect of microtubule inhibitors on malignant invasion *in vitro*. J Nat Cancer Inst 61:787–792.

Markert CL, GE Seidel Jr. 1981. Parthenogenesis, identical twins, and cloning in mammals. In: BG Brackett, GE Seidel, SM Seidel (eds) New Technologies in Animal Breeding. New York, Academic Press, pp 181–200.

Marx JL. 1979. Crown Gall Disease: Nature as genetic engineer. Science 203:254–255.

McClendon JF. 1907. Experiments on the eggs of Chaetopterus and Asterias in which the chromatin was removed. Biol Bull (Woods Hole, Mass) 12:141–145.

McClendon JF. 1908. The segmentation of eggs of Asterias forbesii deprived of chromatin. Arch Entwicklungsmech Org 26:662–668.

McClendon JF. 1910. The development of isolated blastomeres of the frog's egg. Am J Anat 10:425–430.

McGrath J, D Solter. 1983. Nuclear transplantation in mouse embryos. J Exp Zool 228:355–362.

McGrath J, D Solter. 1984a. Maternal T^{hp} lethality in the mouse is a nuclear, not cytoplasmic, defect. Nature 308:550–551.

McGrath J, D Solter. 1984b. Inability of mouse blastomere nuclei transferred to enucleated zygotes to support development in vitro. Science 226:1317–1319.

McKinnell RG. 1960. Transplantation of *Rana pipiens* (Kandiyohi dominant mutant) nuclei to *R. pipiens* cytoplasm. Am Nat 94:187–188.

McKinnell RG. 1962. Intraspecific nuclear transplantation in frogs. J Hered 53:199–207.

McKinnell RG. 1964. Expression of the Kandiyohi gene in triploid frogs produced by nuclear transplantation. Genetics 49:895–903.

McKinnell RG. 1965. Incidence and histology of renal tumors of leopard frogs from the north central states. Ann NY Acad Sci 126:85–98.

McKinnell RG. 1972. Nuclear transfer in *Xenopus* and *Rana* compared. In: R Harris, P Allin, D Viza (eds) Cell Differentiation. Copenhagen, Munksgaard, pp 61–64.

McKinnell RG. 1978. Cloning: Nuclear transplantation in amphibia. Minneapolis, University of Minnesota Press.

McKinnell RG. 1979. The pluripotential genome of the frog renal tumor cell as revealed by nuclear transplantation. Int Rev Cytol, Suppl 9:179–188.

McKinnell RG. 1981. Amphibian nuclear transplantation: State of the art. In: B Brackett, GE Seidel, SM Seidel (eds) New Technologies in Animal Breeding. New York, Academic Press, pp 163–180.

McKinnell RG. 1984. Lucke tumor of frogs. In: GL Hoff, FL Frye, ER Jacobson (eds) Diseases of Amphibians and Reptiles. New York, Plenum Press, pp 581-605.

McKinnell RG, WP Cunningham. 1982. Herpesviruses in metastatic Lucké renal adenocarcinoma. Differentiation 22:41–46.

McKinnell RG, DC Dapkus. 1973. The distribution of burnsi and Kandiyohi frogs in Minnesota and contiguous states. Am Zool 13:81-84.

McKinnell RG, BA Deggins, DD Labat. 1969a. Transplantation of pluripotential nuclei from triploid frog tumors. Science 165:394–396.

McKinnell RG, VL Ellis. 1972. Epidemiology of the frog renal tumour and the significance of tumour nuclear transplantation studies to a viral aetiology of the tumour—a review. In: PM Biggs, G de Thé, LN Payne (eds) Oncogenesis and Herpesviruses. Lyon, Int Agency Res Cancer Sci Publ 2:183–197.

McKinnell RG, E Gorham, FB Martin, JW Schaad. 1979. Reduced prevalence of the Lucké renal adenocarcinoma in populations of *Rana pipiens* in Minnesota. J Nat Cancer Inst 63:821–824.

McKinnell RG, E Gorham, FB Martin. 1980. Continued diminished prevalence of the Lucké renal adenocarcinoma in Minnesota leopard frogs. Am Mid Nat 104:402–404.

McKinnell RG, MF Mims, LA Reed. 1969b. Laser ablation of maternal chromosomes in eggs of *Rana pipiens*. Z Zellforsch Mikrosk Anat 93:30–35.

McKinnell RG, DJ Picciano, RE Krieg. 1976a. Fertilization and development of frog eggs after repeated spermiation induced by human chorionic gonadotropin. Lab Anim Sci 26:932–935.

McKinnell RG, LM Steven Jr, DD Labat. 1976b. Frog renal tumors are composed of Stroma, vascular elements, and epithelial cells: What type nucleus programs for tadpoles with the cloning procedure? In: N Müller-Bérat, C Rosenfeld, D Tarin, D Viza (eds) Progress in differentiation research. Amsterdam, North Holland, pp 319–330.

McKinnell RG, D Tarin. 1984. Temperature-dependent metastasis of the Lucke renal carcinoma and its significance for studies on mechanisms of metastasis. Cancer Metastasis Rev 3:373-386.

McKinnell RG, KS Tweedell. 1970. Induction of renal tumors in triploid leopard frogs. J Nat Cancer Inst 44:1161–1166.

McLaren A, D Michie. 1956. Studies on the transfer of fertilized mouse eggs to uterine foster-mothers. J Exp Biol 33:394–416.

Menkin MF, J Rock. 1948. *In vitro* fertilization and cleavage of human ovarian eggs. Am J Obstet Gynecol 55:440–452.

Mintz B. 1971. Allophenic mice of multi-embryo origin. In: JC Daniels Jr (ed) methods in Mammalian Embryology. San Francisco, Freeman, pp 186–214.

Miyada S. 1960. Studies on haploid frogs. J Sci Hiroshima Univ, Ser B, Div 1

19:1-56.

Modlinski JA. 1981. The fate of inner cell mass and trophectoderm transplanted to fertilized mouse eggs. Nature 292:342-343.

Moler TL, SE Donahue, GB Anderson. 1979. A simple technique for nonsurgical embryo transfer in mice. Lab Anim Sci 29:353-356.

Moore JA. 1955. Abnormal combinations of nuclear and cytoplasmic systems in frogs and toads. Adv Genet 7:139-182.

Mortiz C. 1983. Parthenogenesis in the endemic Australian lizard *Heteronotia binoei* (Gekkonidae). Science 220:735-737.

Morse D, RC Dailey, J Bunn. 1976. Prehistoric multiple myeloma. In: S Jarcho (ed) Essays on the History of Medicine. New York, Science History Publications, pp 413-424.

Moustafa LA, J Hahn. 1978. Experimentelle erzeugung von identischen mäusezwillingen. Dtsch Tieraerztl Wochenschr 85:242-244.

Muggleton-Harris AL. 1979. Reassembly of cellular components for the study of aging and finite life span. Int Rev Cytol, Suppl 9:279-301.

Nayudu PL, A Lopata, PCS Leung, WIH Johnston. 1983. Current problems in human in vitro fertilization and embryo implantation. J Exp Zool 228:203-213.

Okada Y. 1958. The fusion of Ehrlich's tumor cells caused by HVJ virus *in vitro*. Biken's J 1:103-110.

Old RW, SB Primrose. 1980. Principles of Gene Manipulation: An Introduction to Genetic Engineering. Oxford, England, Blackwell Scientific Publications.

Ozil JP. 1983. Production of identical twins by bisection of blastocysts in the cow. J Reprod Fertil 69:463-468.

Parmenter CL. 1933. Haploid, diploid, triploid, and tetraploid chromosome numbers and their origin in parthenogenetically developed larvae and frogs of *Rana pipiens* and *Rana palustris*. J Exp Zool 66:409-453.

Phillip M, D Hermoni, G Potashnik. 1983. Comparison of post-thaw sperm motility after freezing in liquid nitrogen with protective media of either glycerol or glycerol-egg-yolk-citrate. Int J Fertil 28:156-160.

Pierce GB, R Shikes, LM Fink. 1978. Cancer, A Problem of Developmental Biology, Englewood Cliffs, New Jersey, Prentice-Hall.

Porter KR. 1939. Androgenetic development of the egg of *Rana pipiens*. Biol Bull (Woods Hole, Mass) 77:233-257.

Ramsey P. 1970. Fabricated Man: The Ethics of Genetic Control. New Haven, Yale University Press.

Research Group of Cytogenetics, Institute of Zoology, Academy Sinica; Research group of Somatic Cell genetics, Institute of Hydrobiology, Academy Sinica; Research group of Nuclear Transplantation, Chang Jiang Fisheries Research Institute, State Fisheries General Board (1980) Sci Sin (Engl Ed) 23:517-525.

Rock J, MF Menkin. 1944. *In vitro* fertilization and cleavage of human ovarian eggs. Science 100:105-107.

Rollins LA, RG McKinnell. 1980. The influence of glucocorticoids in survival and growth of allografted tumors in the anterior eye chamber of leopard frogs. Develop Comp Immunol 4:283-294.

Rollins-Smith LA, N Cohen. 1982. Effect of thymectomy on development of Lucké renal adenocarcinoma in virus-infected leopard frog tadpoles. J Nat Cancer Inst 68:133-138.

Rorvik DM. 1978. In His Image: The Cloning of a Man. Philadelphia, JV Lippincott.

Roth J, D LeRoith, J Shiloach, JL Rosenweig, MA Lesniak, J Havrankova. 1982. The evolutionary origins of hormones, neurotransmitters, and other extracellular chemical messengers. N Engl J Med 306:523-527.

Roux W. 1888. Beiträge zur Entwickelungsmechanik des Embryo. Ueber die Künstliche Hervorbringung halber Embryonen durch Zerstörung einer der beiden ersten Furchungskugeln, sowie über die Nachentwickelung (Postgeneration) der Fehlenden Körperhälfte. Virchow's Arch Pathol Anat Physiol 114:113–153; Resultate 289–291. English translation by H Laufer. In: BH Willier, JM Oppenheimer (eds) Foundations of experimental embryology. Englewood Cliffs, New Jersey, Prentice-Hall, pp 2–37.

Ruud G. 1925. Die Entwicklung isolierter Keimfragmente frühester Stadien von *Triton taeniatus*. Wilhelm Roux Arch Entwicklungsmech Org 105:209–293.

Sachs MI, E Anderson. 1970. A cytological study of artificial parthenogenesis in the sea urchin *Arbacia punctulata*. J Cell Biol 47:140–158.

Sanyal MK, F Naftolin. 1983. In vitro development of the mammalian embryo. J Exp Zool 228:235–251.

Sawicki JA, T Magnuson, CJ Epstein, 1981. Evidence for expression of the paternal genome in the two-cell mouse embryo. Nature 294:450–451.

Schneider U, RR Maurer. 1983. Factors affecting survival of frozen-thawed mouse embryos. Biol Reprod 29:121–128.

Seidel GE Jr. 1981. Superovulation and embryo transfer in cattle. Science 211:351–358.

Seidel GE Jr. 1983. Production of genetically identical sets of mammals: Cloning? J Exp Zool 228:347–354.

Seppanen ED, RG McKinnell, D Tarin, LA Rollins-Smith, W Hanson. 1984. Temperature-dependent dissociation of Lucké renal adenocarcinoma cells. Differentiation. 26:227–230.

Simpson NS, RG McKinnell. 1964. The burnsi gene as a nuclear marker for transplantation experiments in frogs. J Cell Biol 23:371–375.

Solter D. 1981. Gene transfer in mammalian cells. In: BG Brackett, GE Seidel, SM Seidel (eds) New Technologies in Animal Breeding. New York, Academic Press, pp 201–218.

Spemann H. 1938. Embryonic development and induction. New Haven, Conn, Yale University Press.

Staats, J. 1980. Standardized nomenclature for inbred strains of mice: Seventh Listing. Cancer Res 40:2083–2128.

Stebbins GL Jr. 1950. Variation and Evolution in Plants. New York, Columbia University Press.

Steptoe PC, RG Edwards. 1978. Birth after the reimplantation of a human embryo. Lancet 2:366.

Stevens, LC. 1978. Totipotent cells of parthenogenetic origin in a chimeric mouse. Nature 276:266–267.

Stevens LC. 1980. Teratocarcinogenesis and spontaneous parthenogenesis in mice. In: RG McKinnell, MA Di Berardino, M Blumenfeld, RD Bergad (eds) Differentiation and Neoplasia. Berlin, Springer-Verlag, pp 265–274.

Steward FC. 1970. Totipotency, variation and clonal development of cultured cells. Endeavor 29:117–124.

Surani MAH, SC Barton, ML Norris. 1984. Development of reconstituted mouse eggs suggests imprinting of the genome during gametogenesis. Nature 308:548–550.

Tarin D, RG McKinnell, GW Nace. 1984. Artificially induced metastasis by cells from spontaneous Lucké renal adenocarcinoma. Invasion Metastasis 4:198–208.

Tarin D, JE Price. 1979. Metastatic colonization potential of primary tumour cells in mice. Br J Cancer 39:740–754.

Trounson AO. 1982. Current perspectives on *in vitro* fertilization and embryo transfer. Clin Reprod Fertil 1:55–65.

Trounson A, A Conti. 1982. Research in human *in vitro* fertilization and embryo transfer. Br Med J 285:244–248.

Trounson AO, JF Leeton, C Wood, J Webb, J Wood. 1981. Pregnancies in humans by fertilization *in vitro* and embryo transfer in the controlled ovulatory cycle. Science 212:681–682.

Tsunoda Y, A McLaren. 1983. Effect of various procedures on the viability of mouse embryos containing half the normal number of blastomeres. J Reprod Fertil 69:315–322.

Van Vleck, LD. 1981. Potential genetic impact of artificial insemination, sex selection, embryo transfer, cloning, and selfing in Dairy cattle. In: BG Brackett, GE Seidel, SM Seidel (eds) New Technologies in Animal Breeding. New York, Academic Press, pp 221–242.

Volpe EP. 1980. The amphibian embryo in transplantation immunity. New York, S Karger Publishers.

Volpe EP, RG McKinnell. 1966. Successful tissue transplantation in frogs produced by nuclear transfer. J Hered 57:167–174.

Watkins JF. 1973. Cell fusion in the study of tumor cells. In: Seventh National Cancer Conference Proceedings. Philadelphia, Lippincott, pp 61–63.

Watson JD. 1971. Moving toward the clonal man. Is this what we want? The Atlantic 227(5):50–53.

Watson JD, J Tooze, D. Kurtz. 1983. Recombinant DNA: A short course. New York, Scientific American Books.

Weismann A. 1892. Das Keimplasma. Eine Theorie der Verebung. Jena: G. Fischer. English translation by WN Parker, H Rönnfeldt. 1915. New York, Charles Scribner's Sons.

Weiss MC. 1980. The analysis of cell differentiation by hybridization of somatic cells. In: RG McKinnell, MA Di Berardino, M. Blumenfeld, RD Bergad (eds) Differentiation and Neoplasia. Berlin/Heidelberg, Springer-Verlag, pp 87–92.

Willadsen SM. 1979. A method for culture of micromanipulated sheep embryos and its use to produce monozygotic twins. Nature 277:298–300.

Willadsen SM. 1981. The developmental capacity of blastomeres from 4- and 8-cell sheep embryos. J Embryol Exp Morphol 65:165–172.

Willadsen SM, C Polge. 1981. Attempts to produce monozygotic quadruplets in cattle by blastomere separation. Vet Rec 108:211–213.

Wood C, A Trounson, J Leeton, JM Talbot, B Buttery, J Webb, J Wood, D Jessup. 1981. A clinical assessment of nine pregnancies obtained by *in vitro* fertilization and embryo transfer. Fertil Steril 35:502–508.

Yadav NS, K Postle, RK Saiki, MF Thomashow, M-D Chilton. 1980. T-DNA of a crown gall teratoma is covalently joined to host plant DNA. Nature 287:458–461.

Yudin AL. 1979. Nuclear transplantation studies in *Amoeba proteus*. Int Rev Cytol, Suppl 9:63–100.

Glossary

Adenocarcinoma. A malignant tumor of glandular tissue.

Agar. A gel-like substance frequently used with culture media as a platform for bacterial growth; can be used in a culture dish as a semi-solid platform for nuclear transplantation.

Agrobacterium tumefaciens. A species of bacteria involved in crown-gall tumor induction.

Amoeba. A single-celled animal, a protozoan.

Amphibian. A member of a group of cold-blooded vertebrates that includes frogs, toads, and salamanders.

Androgenesis. Development of an individual under the control of paternal genes only.

Apomixis. Reproduction without fertilization; see Clone.

Bacterial plasmid. DNA of a bacterium that is *not* part of the chromosome of the cell.

Blastomere. A large cell resulting from the division of a fertilized egg.

Blastula. A young embryo—before the start of organ formation—in the form of a ball of cells; the cells are known as blastomeres.

Burnsi. A genetic variant of the common leopard frog, *Rana pipiens*, characterized by the absence of spots on the back.

Capillary tube. A hollow glass of small internal diameter, used for construction of nuclear-transfer micropipettes.

Centrifugation. A procedure used to separate substances of differing density by spinning about an axis.

Chorionic gonadotropin. A hormone used for, among other things, the release of sex cells.

Chromosome. A structure that can be observed in dividing cells which contains a linear arrangement of genes; chromosomes are attached to the spindle during mitosis.

Cleavage. Division of a fertilized egg into daughter cells.

Clone. A group of genetically identical organisms produced without sexual reproduction; an individual produced asexually; see apomixis.

Colcemid. A drug that permits an accumulation of dividing cells, thus making enhanced chromosome analysis possible.

Crown gall. A tumor of plants caused by a plasmid transmitted by the bacterium *Agrobacterium tumefaciens*.

Cumulus oophorous. A portion of the ovarian follicle that contains the growing egg; see follicle.

Cytochalasin B. A drug obtained from a mold that affects the cytoskeleton of living cells.

Cytoplasm. The non-nuclear portion of a cell.

Differentiation. The process by which various cell types develop from a zygote; the evolution of cellular differences.

Diploid individual. An ordinary individual with a full complement of genes from both a maternal and paternal parent.

DNA (Deoxyribonucleic acid). The substance of which genes are constructed.

Drosophila melanogaster. A small fruit fly that has been used extensively in genetic research.

E. coli (Escherichia coli). A common bacterium of the gut.

Ectopic. Out of its ordinary place; when used with "pregnancy," it refers to a fetus growing somewhere other than the uterus.

Electrophoresis. A procedure for the separation of a mixture of substances with differing electric charges and molecular weight in an electric field.

Embryology. The study of developing organisms. Typically, animal embryology encompasses the period of development from fertilization until birth.

Enucleation. Removal of the nucleus from a cell. This can be accomplished by certain chemicals, irradiation, or microsurgery.

Enzyme. A protein catalyst that enhances the rate of a biological reaction.

Eutheria. Mammals with a placenta. Includes most well-known mammals; does *not* include protheria (egg-laying mammals) or metatheria (kangaroos and opossums).

Follicle. The portion of the ovary that contains a growing egg (the oocyte) as well as other cells which nourish and protect the oocyte. Cumulus cells are follicle cells that surround the oocyte.

Gamete. A cell (egg or sperm) specialized for sexual reproduction.

Gastrula. A young embryo composed of ectoderm, mesoderm, and endoderm; an embryonic stage, following the blastula stage, characterized by the onset of cell differentiation.

Genome. A complete set of genes characteristic of a species.

Haploid individual. An individual with genes provided by only *one* parent; see Androgenesis.

Hyaluronidase. An enzyme that catalyzes the breakdown of substances, such as hyaluronic acid, which are found in connective tissue.

Hybrid. Progeny of parents of different genetic stock; also, something of composite origin, as in a "nucleo-cytoplasmic" hybrid.

Immunobiology. The study of immune reactions which includes responses to grafted foreign tissue.

In vitro **fertilization**. The union of a sperm with an egg in a laboratory container. Laboratory dishes were formerly made of glass, hence *in vitro* refers to fertilization in laboratory glassware.

Isogenic. Identical in genetic constitution. Nuclear transplant frogs from a common embryonic nuclear donor are isogenic.

Kandiyohi. A genetic variant of the common leopard frog characterized by a mottled pigment pattern; named for a Minnesota county which was one source of frogs of this type.

Laparoscope. An optical device for viewing the interior of the abdomen through a small incision. Useful for visualizing mature follicles in the procedure for removing oocytes from humans.

Lethal hybrid. Nonviable progeny of parents of different genetic stock; for example, when bullfrog sperm is used to fertilize eggs of the leopard frog, the embryos die at an early stage of development.

Luteinizing hormone. A pituitary hormone that is involved in ovarian follicle maturation and ovulation.

Metamorphosis. Altered anatomy and physiology associated with a changed life style; for example, the transformation from aquatic tadpole to terrestrial frog.

Metastasis. Translocation of a disease from its primary site in a body to a distant site(s) in the same body; used to describe the growth of multiple secondary cancer colonies after dissemination of cells from the primary malignant tumor.

Micromanipulation. Precise movement of something (generally a scientific instrument like a micropipette) on a very small scale.

Micropipette. A glass tube of small internal diameter that tapers to a sharp point on one end; used in nuclear transplantation.

Mitosis. Division of a nucleus with distribution of an equal complement of chromosomes to each daughter nucleus.

Morphology. The science of form and structure of plants or animals.

Multiple myeloma. A bone cancer.

Mutant. A changed individual resulting from an alteration in DNA structure.

Nuclear transplantation. The procedure for placing a nucleus into appropriate cytoplasm; used in the production of cloned frogs.

Nucleolus. An RNA-containing structure in the nucleus of most cells.

Nucleus. A structure found in most kinds of cells that contains the genetic material DNA as well as an RNA-containing smaller body known as the nucleolus.

Oocyte. An unfertilized ovum; the primary oocyte grows in the ovarian follicle, it becomes a secondary oocyte at ovulation, and a mature egg after fertilization.

Oviduct. A tube that originates at the ovary and is confluent with the uterus. The ovulated oocyte is transported through the oviduct to the uterus. The site of fertilization in mammals.

Ovum. Egg.

Parthenogenesis. Development of an unfertilized egg.

Phage. A virus that replicates in a bacterium (short for bacteriophage).

Pituitary gland. Organ located in the floor of the skull that produces hormones which control the activities of other glands; also known as the hypophysis.

Plasmid DNA. See Bacterial plasmid.

Progesterone. A female sex hormone produced by the ovary.

Pronucleus The haploid nucleus produced from the head of a sperm after it fertilizes an egg, or the haploid nucleus of the mature egg. A diploid zygote nucleus is produced from the union of male and female pronuclei.

Protoplast. A cell, which ordinarily has a cell wall, deprived of that cell wall.

Rana catesbieana. The American bull frog.

Rana esculenta. The edible frog of Europe.

Rana pipiens. The North American leopard frog.

Recombinant DNA. The assembly of DNA from different species; generally bacteria! plasmid DNA is linked with nonbacterial DNA, which is inserted into suitable bacterial hosts where the hybrid DNA is replicated.

Restriction endonuclease. Enzymes produced by many bacterial species that provide protection from foreign DNA molecules by digesting that DNA.

Sea urchin. A marine invertebrate related to the sea star.

Sendai virus. A virus that has proved to be of value in causing cells to fuse with one another.

Senescence. Aging.

Surrogate mother. A female mammal that is host (in its uterus) to an embryo that is the biological offspring of another female.

Test-tube baby. An individual who results from *in vitro* fertilization; an egg is fertilized in a laboratory dish and the resulting zygote is inserted into the uterus of an appropriate female, where it develops as does any ordinary fetus.

Tetraploid. A cell with twice the amount of DNA of ordinary body cells—or four times as much DNA as is found in a mature sperm.

Thymus gland. A lymphoid organ that is important in the immune system; thymectomy results in diminished capacity to reject foreign-tissue grafts.

Totipotency. The condition of a nucleus having all genes necessary for normal development and the capacity for their expression to result in a complete organism.

Triploid. Three times the amount of DNA contained in a mature sperm.

Trophectoderm. The cells of an early mammalian embryo that are involved in attachment to the uterus and nourishment of the embryo proper.

Xenopus laevis. The South African clawed toad.

Zona pellucida. A translucent protective noncellular layer that surrounds the oocyte and early cleavage stages of mammals.

Zygote. A fertilized egg.

Index

Made in the USA
Monee, IL
07 July 2026

56552244R00079